T0173088

Judith R. Ahronheim
Editor

High-Level Subject Access Tools and Techniques in Internet Cataloging

High-Level Subject Access Tools and Techniques in Internet Cataloging has been co-published simultaneously as *Journal of Internet Cataloging*, Volume 5, Number 4 2002.

Pre-publication REVIEWS, COMMENTARIES, EVALUATIONS . . .

"AN EXCELLENT OVERVIEW of the ways both traditional and innovative approaches to subject organization can be applied to electronic resources. This anthology brings the fundamental issue of intellectual access to electronic library materials down to earth where readers can readily understand its problems. The book FURNISHES BOTH THEORETICAL BACKGROUND AND NUTS- AND-BOLTS METHODS. . . . HIGHLY RECOMMENDED for library directors, library systems designers, cataloging managers, managers of electronic resources, and librarians in general."

Sheila S. Intner, DLS
Professor Emerita
Simmons College Graduate School of Library & Information Science at Mount Holyoke College

High-Level Subject Access Tools and Techniques in Internet Cataloging

High-Level Subject Access Tools and Techniques in Internet Cataloging has been co-published simultaneously as *Journal of Internet Cataloging,* Volume 5, Number 4 2002.

The *Journal of Internet Cataloging* Monographic "Separates"

Below is a list of "separates," which in serials librarianship means a special issue simultaneously published as a special journal issue or double-issue *and* as a "separate" hardbound monograph. (This is a format which we also call a "DocuSerial.")

"Separates" are published because specialized libraries or professionals may wish to purchase a specific thematic issue by itself in a format which can be separately cataloged and shelved, as opposed to purchasing the journal on an on-going basis. Faculty members may also more easily consider a "separate" for classroom adoption.

"Separates" are carefully classified separately with the major book jobbers so that the journal tie-in can be noted on new book order slips to avoid duplicate purchasing.

You may wish to visit Haworth's Website at . . .

http://www.HaworthPress.com

. . . to search our online catalog for complete tables of contents of these separates and related publications.

You may also call 1-800-HAWORTH (outside US/Canada: 607-722-5857), or Fax 1-800-895-0582 (outside US/Canada: 607-771-0012), or e-mail at:

getinfo@haworthpressinc.com

High-Level Subject Access Tools and Techniques in Internet Cataloging, edited by Judith R. Ahronheim (Vol. 5, No. 4, 2002). *Examines a state-of-the-art cross-section of research data and practical strategies for developing a high-level browse interface.*

Encoded Archival Description on the Internet, edited by Daniel V. Pitti, CPhil, MLIS, and Wendy M. Duff, PhD (Vol. 4, No. 3/4, 2001). *"A broad overview of EAD: how it came to be, what it does, where it fits with other descriptive standards, and what its future might hold. Thought-provoking articles explore EAD's potential impact on reference service, archival information systems, and museum descriptive practices–areas that have not yet been fully explored or exploited by the archival community." (Kris Kiesling, MILS, Head, Department of Manuscripts and Archives, Harry Ranson Humanities Research Center, University of Texas at Austin)*

CORC: New Tools and Possibilities for Cooperative Electronic Resource Description, edited by Karen Calhoun, MS, MBA, and John J. Riemer, MLS (Vol. 4, No. 1/2, 2001). *Examines the nuts-and-bolts practical matters of making a cataloging system work in the Internet environment, where information objects are electronic, transient, and numerous.*

Metadata and Organizing Educational Resources on the Internet, edited by Jane Greenberg, PhD (Vol. 3, No. 1/2/3, 2000). *"A timely and essential reference. . . . A compilation of important issues and views . . . provides the reader with a balanced and practical presentation of empirical case studies and theoretical elaboration." (John Mason, Co-Chair, Dublin Core Education Working Group, and Technical Director, Education, Au LTD [Education Network Australia])*

Internet Searching and Indexing: The Subject Approach, edited by Alan R. Thomas, MA, and James R. Shearer, MA (Vol. 2, No. 3/4, 2000). *This handy guide examines the tools and procedures available now and for the future that will help librarians, students, and patrons search the Internet more systematically, and also discusses how Internet pages can be modified to facilitate easier and efficient searches.*

High-Level Subject Access Tools and Techniques in Internet Cataloging

Judith R. Ahronheim
Editor

High-Level Subject Access Tools and Techniques in Internet Cataloging has been co-published simultaneously as *Journal of Internet Cataloging,* Volume 5, Number 4 2002.

The Haworth Information Press
An Imprint of
The Haworth Press, Inc.
New York • London • Oxford

Published by

The Haworth Information Press®, 10 Alice Street, Binghamton, NY 13904-1580 USA

The Haworth Information Press® is an imprint of The Haworth Press, Inc., 10 Alice Street, Binghamton, NY 13904-1580 USA.

High-Level Subject Access Tools and Techniques in Internet Cataloging has been co-published simultaneously as *Journal of Internet Cataloging,* Volume 5, Number 4 2002.

Cover design by Lora Wiggins.

Library of Congress Cataloging-in-Publication Data

High-level subject access tools and techniques in Internet cataloging / Judith R. Ahronheim, editor.
     p. cm.
    Co-published simultaneously as Journal of Internet cataloging, v. 5, no. 4, 2002.
    Includes bibliographical references and index.
    ISBN 0-7890-2024-6 (alk. paper) – ISBN 0-7890-2025-4 (pbk : alk. paper)
    1. World Wide Web–Subject access. 2. Cataloging of computer network resources. I. Ahronheim, Judith R. II. Journal of Internet cataloging.
ZA4232 .H54 2002
025. 04–dc21

                                2002151190

Indexing, Abstracting & Website/Internet Coverage

This section provides you with a list of major indexing & abstracting services. That is to say, each service began covering this periodical during the year noted in the right column. Most Websites which are listed below have indicated that they will either post, disseminate, compile, archive, cite or alert their own Website users with research-based content from this work. (This list is as current as the copyright date of this publication.)

Abstracting, Website/Indexing Coverage . . . . Year When Coverage Began

- *AGRICOLA Database <www.natl.usda.gov/ag98>* **1997**

- *Applied Social Sciences Index & Abstracts (ASSIA) (Online: ASSI via Data-Star) (CDRom: ASSIA Plus) <www.csa.com>* **1997**

- *CNPIEC Reference Guide: Chinese National Directory of Foreign Periodicals* **1997**

- *Combined Health Information Database (CHID)* **1997**

- *Computer Literature Index* **1997**

- *Current Cites [Digital Libraries] [Electronic Publishing] [Multimedia & Hypermedia] [Networks & Networking] [General]* **1998**

- *FINDEX <www.publist.com>* **1999**

- *FRANCIS. INIST/CNRS <www.inist.fr>* **1999**

(continued)

- *IBZ International Bibliography of Periodical Literature*
  <www.saur.de> **1999**

- *Index to Periodical Articles Related to Law* **1997**

- *Information Science Abstracts <www.infotoday.com>* **1997**

- *INSPEC <www.iee.org.uk/publish/>* **1999**

- *Internet & Personal Computing Abstracts can be found online
  at: DIALOG, File 233; HRIN & OCLC and on Internet
  at: Cambridge Scientific Abstracts; Dialog Web & OCLC
  <www.infotoday.com/mca/default.htm>* **1999**

- *Konyvtari Figyelo (Library Review)* **1997**

- *Library & Information Science Abstracts (LISA)*
  <www.csa.com> **1997**

- *Library and Information Science Annual (LISCA)*
  <www.lu.com> **1999**

- *Management & Marketing Abstracts* **2000**

- *MLA International Bibliography <www.mla.org>* **1997**

- *OCLC Public Affairs Information Service*
  <www.pais.org> **1999**

- *PASCAL, c/o Institut de l'Information Scientifique et
  Technique. Cross-disciplinary electronic database
  covering the fields of science, technology & medicine.
  Also available on CD-ROM, and can generate customized
  retrospective searches <www.inist.fr>* **1998**

- *Referativnyi Zhurnal (Abstracts Journal of the All-Russian
  Institute of Scientific and Technical Information–
  in Russian)* **1997**

- *Social Work Abstracts <www.silverplatter.com/catalog/swab.htm>* .. **1997**

- *SwetsNet <www.swetsnet.com>* **2001**

- *World Publishing Monitor* **2000**

(continued)

Special Bibliographic Notes related to special journal issues
(separates) and indexing/abstracting:

- indexing/abstracting services in this list will also cover material in any "separate" that is co-published simultaneously with Haworth's special thematic journal issue or DocuSerial. Indexing/abstracting usually covers material at the article/chapter level.
- monographic co-editions are intended for either non-subscribers or libraries which intend to purchase a second copy for their circulating collections.
- monographic co-editions are reported to all jobbers/wholesalers/approval plans. The source journal is listed as the "series" to assist the prevention of duplicate purchasing in the same manner utilized for books-in-series.
- to facilitate user/access services all indexing/abstracting services are encouraged to utilize the co-indexing entry note indicated at the bottom of the first page of each article/chapter/contribution.
- this is intended to assist a library user of any reference tool (whether print, electronic, online, or CD-ROM) to locate the monographic version if the library has purchased this version but not a subscription to the source journal.
- individual articles/chapters in any Haworth publication are also available through the Haworth Document Delivery Service (HDDS).

High-Level Subject Access Tools and Techniques in Internet Cataloging

CONTENTS

Introduction   1
Judith R. Ahronheim

Classification Schemes for Internet Resources Revisited   5
Diane Vizine-Goetz

HILCC: A Hierarchical Interface to Library of Congress
   Classification   19
Stephen Paul Davis

University of Washington Libraries Digital Registry   51
Kathleen Forsythe
Steve Shadle

Bridging the Gap Between Materials-Focus
   and Audience-Focus: Providing Subject Categorization
   for Users of Electronic Resources   67
Jonathan Rothman

Competing Vocabularies and "Research Stuff"   81
Keith A. Morgan
Tripp Reade

HILT: Moving Towards Interoperability in Subject
   Terminologies   97
Dennis Nicholson
Gordon Dunsire
Susannah Neill

Index   113

ABOUT THE EDITOR

Judith R. Ahronheim, AMLS, is Metadata Specialist Librarian at the University Library of the University of Michigan and was Head of the Original Cataloging Unit for seven years. She has also served as Reference Librarian at the Lake County Public Library in Indiana. She is active in ALA's Networked Resources and Metadata Committee, has worked on several Digital Library Federation projects, and was the 1998 recipient of ALA's Esther J. Piercy award. She is a co-editor of *Cataloging the Web: Metadata, AACR, and MARC 21* and her other publications include the chapter "Reducing Complexity in Bibliographic Systems" in *Finding Common Ground*, as well as articles appearing in the *Journal of Academic Librarianship, Cataloging & Classification Quarterly*, and *Technicalities*.

Introduction

Judith R. Ahronheim

As the World Wide Web and graphic user interfaces (GUIs) developed in the 90s, libraries began to build gateways for their online resources. These gateways allowed library users to employ the browse, point, and click approach to resource discovery that they had come to expect from online tools. Extremely popular at the end of the 90s, most of these interfaces amounted to little more than hand-constructed lists of links. As online resource content grew, librarians began to organize these lists by topic, but still constructed them by hand. The success of Yahoo! encouraged libraries to continue to mimic this model even as the increasing volume of content to be included and a high rate of URL changes introduced significant maintenance burdens. Today, many libraries offer access to users through a set list of broad topics, sometimes called a high-level browse display. These displays can offer customizable options a la MyYahoo and MyLibrary and may sometimes lead users deeper into a subject hierarchy. Methods for populating these subject categories remain crude and maintenance requires considerable resources. As a result, libraries have begun to look at ways of applying traditional techniques associated with cataloging to these new interfaces. Several goals are involved in these developments. Many hope to be able to reuse data from library catalogs and thus to limit maintenance burdens. Others seek to apply a more standard set of tools and principles to the construction of portals so as to allow greater cooperation among institutions who want to interoperate with one another.

Judith R. Ahronheim, currently Metadata Specialist at the University of Michigan Library, was head of its Original Cataloging Unit for seven years.

[Haworth co-indexing entry note]: "Introduction." Ahronheim, Judith R. Co-published simultaneously in *Journal of Internet Cataloging* (The Haworth Information Press, an imprint of The Haworth Press, Inc.) Vol. 5, No. 4, 2002, pp. 1-3; and: *High-Level Subject Access Tools and Techniques in Internet Cataloging* (ed: Judith R. Ahronheim) The Haworth Information Press, an imprint of The Haworth Press, Inc., 2002, pp. 1-3. Single or multiple copies of this article are available for a fee from The Haworth Document Delivery Service [1-800-HAWORTH, 9:00 a.m. - 5:00 p.m. (EST). E-mail address: getinfo@haworthpressinc.com].

To date, very little has been written about how high-level browsing vocabularies are developed and used for library purposes. Kirsten Risden's research is an early example of testing methods for determining the usability of categories at a high level.[1] Her methods, however have not produced additional studies that could guide libraries in choosing topics for use at the top of subject trees. As Alan Wheatley notes, "Though individual subject trees, especially Yahoo, are frequently described in articles on Internet search engines, it is rare for a number of them to be described and compared and the prime concern of authors who do so has usually been the information content of subject trees and not their broad structure or classified organization."[2] His article offers a survey of six commercial Web sites and notes similarities in the distribution of topics at the upper levels. Little other information of this type has been published. Diane Vizine-Goetz's article in this book continues this exploration of subject trees by comparing those of two commercial sites to the structure inherent in the Dewey Classification. The implication here is that if Dewey's structure is a good enough mirror of successful subject trees, it could be used to build and populate trees using Dewey classification extant in catalogs such as WorldCat. We have just begun to learn about the forces shaping high-level browse and know little about the general characteristics of such tools.

Others, though, have shared the idea of building access tools based on existing classification data. In this book, Stephen Davis describes Columbia's efforts to build an automatically generated browsable display based on Library of Congress Classification as it occurs in catalog records. Davis' article describes a set of activities that are mirrored in the efforts of other libraries attempting to produce high-level browsing lists. Both his article and those by Kathleen Forsythe and Steve Shadle and by Jonathan Rothman show evidence of emerging patterns in the development of browsing services. All speak of developing a hierarchy of subjects for browsing that is not based in classification, a map that relates data from catalog records to the subject hierarchy and tools for extracting data from a catalog and storing it in a separate database to produce a more flexible display.

Another theme appears in many of the articles in this work. Authors refer to these subject lists as being "task-based" (Forsythe and Shadle) or "user-focused" (Rothman) rather than materials-based. Morgan and Reade's article places the categorizing of resources within this context and examines some of the social issues associated with choosing categories based on the nature and activities of our users. Both this article and Rothman's refer to the political issues involved in selecting disciplines or topics for a browsing service. This political process is brought closest to home in the final article, describing the High Level Thesaurus Project (HILT), in which a group of libraries, archives and museums attempted to find a common method for high level subject access via portal. Here, Nicholson, Dunsire and Neill report on the challenges encountered when representatives from a variety of cultural heritage communities get together to agree upon methods for searching their distributed catalogs. In this instance the goal is a shared search, rather than a browse hierarchy, but the choice of mapping, used here to lead a user from many subject vocabularies to the Dewey Decimal classification as a common denominator, is common to the other articles in the publication and may prove helpful in later developments of browsing tools. In all the articles in this publication we see the confluence of Web-type tools and traditionally maintained library structures. New tools are also being developed that might make use of our knowledge. Work with topic maps[3] is still too new to have produced papers describing library applications, but that will no doubt come in the next few years. Meanwhile, much investigation remains before we can say that any such confluences can provide effective tools for organizing the information universe. The efforts described here are good first steps in testing the hypothesis.

NOTES

1. Kirsten Risden, "Toward Usable Browse Hierarchies for the Web," in *Human Computer Interaction: Ergonomics and User Interfaces*, Hans-Jorg Bull, ed. (1999). Also available on the Web as a pdf file: <http://www.microsoft.com/usability/UEPostings/HCI-kirstenrisden.doc>.
2. Alan Wheatley, "Subject Trees on the Internet," *Journal of Internet Cataloging* 2, no. 3-4 (2000): 135.
3. Steve Pepper's "The Tao of Topic Maps" is a good introduction to the subject. A PDF version can be found at <www.ontopia.net/topicmaps/materials/tao.pdf>.

Classification Schemes
for Internet Resources Revisited

Diane Vizine-Goetz

SUMMARY. Library classification schemes have become increasingly available in electronic form and undergone many enhancements that make them attractive for Web knowledge organization. In fact, library professionals have been quite successful in applying library classification to Internet-based information services in a number of projects, both small and large. Yet, many opportunities remain for improving our general knowledge organization tools and using them in new ways. In this article, the DDC hierarchy structure is compared to the subject trees of Internet directory services in terms categories, hierarchies, and distributions of postings. The schemes are also compared with respect to several general characteristics that support browsing. The findings suggest that the prospects are very good for developing effective DDC-based browsing structures to large collections. *[Article copies available for a fee from The Haworth Document Delivery Service: 1-800-HAWORTH. E-mail address: <getinfo@haworthpressinc.com> Website: <http://www.HaworthPress.com> © 2002 by The Haworth Press, Inc. All rights reserved.]*

KEYWORDS. Browse structure, category structure, DDC, Dewey Decimal Classification, subject tree, Web directory service

---

Diane Vizine-Goetz is Consulting Research Scientist, OCLC Office of Research. She received her PhD in Library and Information Science from Case Western Reserve University in 1983.

The author would like to thank Roger Thompson for analyzing the bibliographic file and matching to the DDC, Joan Mitchell and Julianne Beall for their careful review of the paper and helpful suggestions, and Nick G. for assistance computing the statistics.

[Haworth co-indexing entry note]: "Classification Schemes for Internet Resources Revisited." Vizine-Goetz, Diane. Co-published simultaneously in *Journal of Internet Cataloging* (The Haworth Information Press, an imprint of The Haworth Press, Inc.) Vol. 5, No. 4, 2002, pp. 5-18; and: *High-Level Subject Access Tools and Techniques in Internet Cataloging* (ed: Judith R. Ahronheim) The Haworth Information Press, an imprint of The Haworth Press, Inc., 2002, pp. 5-18. Single or multiple copies of this article are available for a fee from The Haworth Document Delivery Service [1-800-HAWORTH, 9:00 a.m. - 5:00 p.m. (EST). E-mail address: getinfo@haworth pressinc.com].

INTRODUCTION

Approximately 6 years ago I assessed the potential of library classification schemes such as the Dewey Decimal Classification (DDC) and the Library of Congress Classification (LCC) to provide access to Internet resources.[1] An examination of the content of selected categories, structure of hierarchies, and role of notation lead to the following observations:

- Library schemes compare favorably to Internet schemes in terms of overall coverage of topics
- Library schemes have hierarchies that are broad and deep enough to support effective browsing
- Library schemes, which contain class notations, have an advantage over Internet schemes since class numbers can be used to manipulate categories for browsing and retrieval.

That study also identified several improvements that should be made to library schemes (e.g., updated captions, new terminology, and more links to other vocabularies) if they were to become a major tool for providing access to both traditional collections and Internet resources.

Since then, library classification schemes have become increasingly available in electronic form and undergone many enhancements.[2-4] Additionally, library professionals have been quite prolific in applying library classification to Internet-based information services. The number of electronic resources classed by Dewey and LCC has also risen dramatically. In 1996 there were fewer than 1,000 bibliographic records for Internet resources in the WorldCat database.[5] Now there are more than 500,000 records[6] with hundreds of thousands of DDC and LCC class numbers assigned to them.[7] Improved access to Web versions of the DDC and licensing of the database for research purposes have led to several Internet projects that are exploring end user applications of Dewey. The projects serve a range of needs and audiences:

- BIOME,[8] which provides access to Internet resources in the health and life sciences, is aimed at students, academics and practitioners. The DDC is used in two of its subject gateways (Agrifor and Natural World) to support browsing;
- BUBL LINK,[9] run by University of Strathclyde, is an Internet-based information service for the UK higher education community. BUBL presents a directory style interface to Web resources based on the DDC;
- Renardus,[10] a collaborative project involving several European subject gateways, is developing an academic subject gateway to support learning

and research. The DDC is being used to provide a common browsing structure and switching language for the different subject vocabularies used by the project partners.

With the advances have come opportunities for improving our general knowledge organization tools and using them in new ways. For example, the hierarchy structures of library classification schemes are thought to have great potential as browsing structures. They are often compared to the subject trees of Internet directory services, which are now the norm on the Web; however, research is needed to determine if library schemes require adaptations if they are to be used in a similar way to support browsing. In this paper, the DDC scheme and its application in several million bibliographic records are compared to the subject trees of Yahoo! and LookSmart. It is reasonable to compare the DDC to these services since they employ extensive classificatory structures that have been applied to large numbers of resources.[11] Their use of subject categorization is similar in scope and scale to the application of traditional classification schemes in library databases. The schemes are also compared with respect to several general characteristics that support browsing.

LIBRARY CLASSIFICATION SCHEMES AND INTERNET SUBJECT TREES

Background

An article by Alan Wheatley[12] published in 1999 provided the basis for the comparisons in this paper. In his article, Wheatley gives estimates of the number of categories and postings (Web sites, etc.) at various levels of hierarchy for the subject trees of major Internet directory services. He describes several properties of the schemes based on these analyses and on accepted views of the schemes. The following are relevant to this study:

- Subject trees have about 12-20 top level headings
- Top-level headings have almost no postings
- Two-thirds of the postings are concentrated at levels 4 or 5
- Subject trees provide the advantages of browsing without the complexities of traditional bibliographic classification schemes
- Subject trees provide alphabetical and hierarchical access without using the formal term relationships (broader terms, narrower terms, etc.) found in traditional thesauri
- Subject trees can be applied to multilingual collections.

For this paper the DDC was analyzed in a similar way. Statistics for total number of available categories, number of categories by level, distribution of resources across categories and levels, and average category size were computed for the categories in the DDC scheme and the postings in a large file of bibliographic resources. The file consisted of 1.8 million bibliographic records with DDC numbers extracted from the OCLC WorldCat database.[13]

Categories and Hierarchy

Built Numbers

The DDC has been described as a "number-building machine."[14] Built numbers represent categories not specifically named in the schedules. They are created by combining listed topics according to instructions given in the Dewey classification schedules.[15] Although built numbers for many complex topics are included in the DDC, they account for only a fraction of the potential topics that can be represented. The following example may help to illustrate. The built number, 635.91531 for the category, propagating ornamental plants from seeds, is created by combining a class number in the schedules with a part of another schedule number (in this example, the numbers following 631 in 631.2-631.8)

Flowers and ornamental plants

Specific techniques; apparatus, equipment, materials   635.91

Propagation from seeds   531

This is one of the many ways of building class numbers in the DDC. In Figure 1, categories in bold and bold italic type represent schedule numbers and built class numbers that are included in the print and electronic version of the DDC. Categories in normal type represent additional built class numbers, created by librarians, found in the bibliographic file.

Class numbers of the form 6xx were assigned to level 1, 63x to level 2, 635 to level 3, 635.9 to level 4, and so on. Category levels were calculated on the length of the DDC notation except for standard subdivisions and built numbers, which received special processing. Standard subdivisions were treated as an implied level of general topics (see 635.90* in Figure 1) under which categories for standard subdivisions were grouped at the next deeper level. The hierarchy level for built numbers was calculated by adding 1 to the level of the base number. Category levels for standard subdivisions and built numbers were calculated in this

FIGURE 1. Expanded Array of Dewey Categories

DDC Class	Category Name	Level
635.9	**Flowers and ornamental plants**	4
635.90*	Standard subdivisions and special topics	5
635.903	Gardening–Dictionaries	6
635.905	Gardening–Periodicals	6
635.90941	Gardening–Great Britain	6
635.90973	Gardening–United States	6
635.91	**Specific techniques; apparatus, equipment, . . .**	5
635.9152	*Nurseries (For plants)–floriculture*	6
635.91521	*Seeds–nursery production*	6
635.91526	*Bulbs (Plants)–nursery production*	6
635.91531	*Seeds–sowing*	6
635.91532	*Bulbs (Plants)–planting*	6
635.91541	*Grafting (Plant propagation)*	6
635.91542	*Pruning*	6
635.91546	*Arbors–floriculture*	6
635.92	**Injuries, diseases, pests**	5

*Not a class in the DDC but indicates the location of standard subdivisions and special topics.

way to group related classes for the purpose of browsing when intermediate levels of hierarchy are not explicitly represented. The categories **635.91531 Flowers and ornamental plants–Seeds–sowing and 635.91532 Flowers and ornamental plants–Bulbs (Plants)–planting** are built number examples. The portion of the hierarchy for specific techniques in agriculture from which these numbers are built consists of the following categories:

631.<u>5</u>	Cultivation and harvesting
631.<u>53</u>	Plant propagation
631.<u>531</u>	**Propagation from seeds (Sowing)**
631.<u>532</u>	**Propagation from bulbs and tubers**

Two of these topics, **631.531 Propagation from seeds**, and **631.532 Propagation from bulbs and tubers,** were combined with the topic **639.91 Flowers** and ornamental plants–Specific techniques to form the built numbers **635.91531 Flowers and ornamental plants–Seeds–sowing** and **635.91532 Flowers and ornamental plants–Bulbs (plants)–planting**. Built numbers for the topics, Cultivation and harvesting and Plant propagation, in the agricultural techniques hierarchy and the topic, Flowers and ornamental plants, did not occur in DDC itself or in the bibliographic records in this study. If levels had been calculated on the length of the notation as they were for non-built numbers, the topic **Seeds–sowing** would have been separated from the topic **Flowers and ornamental plants–Specific techniques** by 2 levels of missing hierarchy (635.915 and 635.9153). Supplying the intermediate levels to complete the hierarchies for built numbers was beyond the scope of this study. The approach employed in this study had the effect of grouping some categories at a higher level than they would have been had the class numbers been fully analyzed. Levels and postings for built vs. non-built numbers can be compared using Tables 1 and 2 and Appendix 2, which contains the statistics for non-built numbers in the DDC. Further research is needed to determine how to code and present hierarchies for built numbers.

Two sets of statistics were computed for categories in the Dewey classification scheme. The first set, called DDC Outline, includes all schedule numbers and built class numbers in the print and electronic versions of the DDC (categories in bold and bold italic in Figure 1). The second set, called the Expanded DDC, includes all class numbers in the first set plus additional built class numbers in the bibliographic file (all categories in Figure 1). The second set, which reflects the impact of number building, presents a more accurate picture of Dewey's ability to represent topics than the set based only on the categories in the print and electronic version of the DDC.

Wheatley's figures for Yahoo! and LookSmart are given in Table 1 along with the distribution of DDC categories by level. The total number of categories in the expanded DDC (44,350) falls midway between the total number of categories in Yahoo! (63,746) and the number in LookSmart (20,670). Despite differences at given category levels, the DDC appears to be both broad enough and deep enough to support browsing in the manner of Internet subject trees. The DDC has the deepest structure with 13 category levels and LookSmart the shallowest with only 6 levels. Yahoo! and DDC provide about the same number of topics at levels 4-7. There are 40,795 for Yahoo! and 39,207 for the Expanded DDC. Yahoo! provides about twice as many topics as the DDC at level 7, although some of the difference is likely due to the way hierarchies for built numbers were computed for the Dewey. It would be interesting nonetheless to learn more about the categories at the lower levels of Yahoo! Are they innovative subarrangements that could be adapted to library classification schemes; or, are

TABLE 1. Distribution of Categories by Level for DDC, Yahoo! and LookSmart

Category Level	DDC Outline	Expanded DDC	Yahoo!	LookSmart
1	10	10	14	13
2	99	99	420	160
3	897	897	3,509	1,681
4	4,415	8,121	7,562	8,604
5	8,574	11,372	11,058	9,191
6	9,986	12,360	6,823	1,021
7	5,590	7,354	15,352	0
8	1,820	2,917	16,632	0
9	535	923	2,376	0
10	93	248	0	0
11	20	36	0	0
12	9	11		
13	0	2		
Total	32,048	44,350	63,746	20,670*

* Wheatley reports a total of 20,675.

they the alphabetical breakdowns referred to by Wheatley? In fact, there is a hidden hierarchy at many levels in Dewey, not represented in Table 1, where subarrangement by instance is preferred. It can be found in categories for computer language, name of TV program, make of car, etc., which are subarranged alphabetically.

Categories and Postings

The DDC hierarchy when viewed as a browsing structure for bibliographic records fits the pattern observed by Wheatley for Internet subject trees where few resources are posted at the broadest category levels. See Table 2, Distribution of resources by category level. The counts of resources by level for Yahoo! and Looksmart are again given alongside the counts for DDC. The DDC's broad-to-specific category structure is easily understood; but it is sometimes criticized for having some categories at lower levels that are as important or more important than categories at higher levels. On the surface the DDC notation would appear to impose constraints on the browsing structure, but in reality it can provide a great deal of flexibility. The notation can be manipulated to elevate whole branches of the tree to higher levels for display. For example, various topics in engineering which occur at levels 4 and 5 of the DDC hierarchy could be presented at level 3 and also at their regular positions in the hierarchy

Technology > Engineering

Aerospace engineering (level 4)

Automotive engineering (level 4)

Chemical engineering

Civil engineering

Hydraulic engineering

Military and nautical engineering

Mining engineering

Nuclear engineering (level 5)

Railroads and roads

Sanitary, municipal, & environmental engineering

Graphical browsing techniques also have the potential to facilitate hierarchy navigation. The Renardus project, mentioned in the Introduction, is experimenting with a graphical display of the DDC that provides an overview of all categories that surround a given category. The typical display shows all categories one level above and two levels below in the hierarchy. This project will eventually provide the DDC browsing structure in some of the languages of the DDC translations (e.g., French and German). The project currently provides access to a multilingual collection through an English language interface.

In spite of efforts to make navigation easier, if a majority of resources are categorized at lower levels, users still might have to browse down several levels of the hierarchy before reaching relevant information. Interestingly, the results for the DDC were very much in line with Wheatley's findings. Like Internet subject trees, approximately two thirds of DDC postings were at level 5 and above. As shown in Table 2, 69.3% of DDC postings were at level 5 and

TABLE 2. Distribution of Resources by Category Level

Category Level	Expanded DDC	%	Yahoo!	%	LookSmart	%
1	0	0.0%	12	0.0%	0	0.0%
2	0	0.0%	11,209	2.5%	243	0.1%
3	220,786	12.2%	79,764	17.7%	30,514	6.6%
4	635,790	35.1%	132,782	29.4%	183,519	39.5%
5	399,097	22.0%	86,114	19.1%	243,391	52.4%
Subtotal		69.3%		68.6%		98.5%
6	326,187	18.0%	43,072	9.5%	7,147	1.5%
7	156,204	8.6%	51,174	11.3%	0	0.0%
8	57,122	3.2%	38,015	8.4%	0	0.0%
9	13,884	0.8%	9,504	2.1%	0	0.0%
10	2,848	0.2%	0	0.0%	0	0.0%
11	598	0.0%	0	0.0%	0	0.0%
12	126	0.0%	0	0.0%	0	0.0%
13	49	0.0%				
Total	1,812,691	*100.1%	451,646	100.0%	464,814	100.0%
Ave. Items/ Category	41		7.1		22.5	

* Total is greater than 100% due to rounding.

above as compared to 68.6% for Yahoo! This percentage holds even when built class numbers are removed from the DDC statistics. See Appendix 2. Although, Yahoo! and Dewey both have deep hierarchies (a maximum of 7 and 13 levels, respectively) only 13% of DDC categorized items are at level 7 or below as compared to 22% for Yahoo! Because directory services often list Web sites in more than one category and categories at more than one level, users of Yahoo! may not necessarily have to browse as deep as the figures imply.

Collection Size

The figures in Table 2 indicate that category size may be a more significant impediment to effective browsing than depth of hierarchy, especially for a very large collection of resources. The DDC, with 4 times as many resources as the other collections, had the highest average number of items per category at 41. LookSmart followed with 22.5. The average number of items per category by level was also computed for the three schemes (Table 3). Again the DDC had the highest average number of items per category at any level, 246.1 at level 3. The second highest was Yahoo! with 26.7 at level 2. Lists of 2 to 27 items (the low for LookSmart and high for Yahoo!) can be scanned without much difficulty by users, but a list of over 200 items is unmanageable, even if presented across several pages. We know that for the DDC many of the groups are much larger than what is indicated by the averages. This is no doubt true for the subject trees as well. For the DDC collection the largest group, **American fiction–1945-1999**, contained 60,578 items. Appendix 1 contains the top 25 categories by category size in the DDC collection. Although the DDC was not designed to provide a further subarrangement for many of these categories, links to other knowledge structures can provide some options.

For several years the Dewey editors and researchers at OCLC have been mapping Dewey classes and Library of Congress subject headings through a variety of editorial and statistical techniques. About 87,000 mappings have been made. Many of the LCSH are added to the electronic version of the DDC to provide additional indexing vocabulary for topics in the classification. The mapped headings could be used just as well to provide additional subject breakdowns for highly posted classes. For example, hundreds of headings for fictitious characters have been mapped to 813.54:

Durell, Sam (Fictitious character)

Duvall, Cheney (Fictitious character)

Eaton, Jake (Fictitious character)

Eckert, James (Fictitious character)

Edmund (Fictitious character: Jackson)

Eldridge, Louise (Fictitious character)

Elliot Moose (Fictitious character)

Elliott, Maggie (Fictitious character)

Elliott, Scott (Fictitious character)

Ellis, Trade (Fictitious character)

Elminster (Fictitious character)

Epps, Margalo (Fictitious character)

Fairchild, Faith Sibley (Fictitious character)

Fansler, Kate (Fictitous character)

TABLE 3. Average Number of Items per Category by Level

Category Level	Expanded DDC	Yahoo!	LookSmart
1		0.9	0.0
2		26.7	1.5
3	246.1	22.7	18.2
4	78.3	17.6	21.3
5	35.1	7.8	26.5
6	26.4	6.3	7.0
7	21.2	3.3	
8	19.6	2.3	
9	15.0	4.0	
10	11.5		
11	16.6		
12	11.5		
13	24.5		

Numerous other helpful mappings have been made for highly posted categories. When systematic approaches aren't applicable, large categories could be ordered by date, popularity of the resource or some combination of the two.

CONCLUSIONS

The analyses in this paper indicate that the DDC's category structure and Internet subject trees have many common properties. Both Internet subject trees and the DDC provide hierarchical and alphabetical access to collections. Both have category structures that can be applied in a multilingual environment. The DDC would seem to have an advantage here since it has been translated into 30 languages. And, like the postings in Internet subject trees, more than two-thirds of the postings of a DDC categorized collection were at level five or above, meaning that the majority of relevant resources are located in the upper third of the DDC's hierarchy structure. Overall, the findings suggest that the prospects are very good for developing effective DDC-based browsing structures to large collections.

NOTES

1. Diane Vizine-Goetz, "Using Library Classification Schemes for Internet Resources." Proceedings of the OCLC Internet Cataloging Colloquium, San Antonio, Texas (January 19, 1996). [Position paper. Colloquium held in conjunction with the American Library Association Midwinter meeting in San Antonio, Texas.] <http://staff.oclc.org/~vizine/Intercat/vizine-goetz.htm>.

2. Joan S. Mitchell, "Dewey Decimal Classification: 125 and still growing." *OCLC Newsletter*, 254 November/December (2001): 27-29. The Dewey Decimal Classification (DDC) system is Copyright 1996-2002 OCLC Online Computer Library Center, Incorporated. Dewey, DDC, and Dewey Decimal Classification are registered trademarks of OCLC.

3. Dawn Lawson, "You've come a long way, Dewey!" *OCLC Newsletter*, 254 November/December (2001): 34-35.

4. Cataloging Distribution Service. Library of Congress. "Classification Web Pilot Test Ended August 31, 2001 . . . Production Service to Start Early 2002." <http://lcweb.loc.gov/cds/classwebpilot.html>.

5. Internet cataloging project database now available. <http://www.oclc.org/oclc/press/950721a.htm>.

6. WorldCat now contains more than 500,000 records for digital resources. <http://www.oclc.org/oclc/press/20011004.shtm>.

7. Diane Vizine-Goetz. Comments. <http://lcweb.loc.gov/catdir/bibcontrol/vizinegoetz_paper.html>.

8. Browse BIOME. <http://biome.ac.uk/browse>.

9. BUBL LINK / 5:15. <http://bubl.ac.uk/link/>.

10. Renardus. <http://www.renardus.org/>.

11. Directory Sizes. <http://www.searchenginewatch.com/reports/directories.html>.

12. Alan Wheatley. "Subject Trees on the Internet: A New Role for Bibliographic Classification," *Journal of Internet Cataloging*, 2:3/4 (2000) 115-141.

13. The records, which were created and input by national bibliographic agencies, were extracted from the WorldCat database in May 2001. See http://www.oclc.org/about/ for more about OCLC and WorldCat.

14. Lois Chan et al., Dewey Decimal Classification: a Practical Guide (Albany, N.Y.: Forest Press, 1996) p. 8.

15. For more information on number building see <http://www.oclc.org/dewey/about/about_the_ddc.htm#building>.

APPENDIX 1

Top 25 Categories by Category Size

Rank	Class	Count	Caption
1	813.54	60,578	American fiction–1945-1999
2	811.54	23,009	American poetry–1945-1999
3	929.20973	22,252	Family histories–United States
4	823.914	21,958	English fiction–1945-1999
5	813.52	9,903	American fiction–1900-1945
6	823.912	9,360	English fiction–1900-1945
7	843.914	8,004	French fiction–1945-1999
8	863	7,283	Spanish fiction
9	823	6,261	English fiction
10	861	5,966	Spanish poetry
11	821.914	5,102	English poetry–1945-1999
12	822.33	4,978	William Shakespeare
13	823.8	4,439	English fiction–1837-1899
14	812.54	4,104	American drama–1945-1999
15	510	4,068	Mathematics
16	641.5	3,678	Cooking
17	841.914	3,625	French poetry–1945-1999
18	821	3,499	English poetry
19	895.135	3,486	Chinese fiction–1912-
20	759.13	3,444	Painting, American
21	811.52	3,141	American poetry–1900-1945
22	843.912	3,134	French fiction–1900-1945
23	248.4	3,029	Christian life and practice
24	081	2,987	Quotations, American
25	833.914	2,943	German fiction–1945-1990

APPENDIX 2

Statistics for DDC Without Built Numbers

Level	Categories	Postings	% Postings/ Level	Cum %	Ave. Postings/ Category
1	10	0	0.0%	0.0%	0.0
2	99	0	0.0%	0.0%	0.0
3	897	220,786	20.9%	20.9%	246.1
4	2,711	232,601	22.1%	43.0%	85.8
5	6,462	272,381	25.8%	68.8%	42.2
6	7,053	204,709	19.4%	88.2%	29.0
7	3,666	94,535	9.0%	97.2%	25.8
8	1,255	23,698	2.2%	99.4%	18.9
9	315	5,051	0.5%	99.9%	16.0
10	72	764	0.1%	100.0%	10.6
11	14	172	0.0%		12.3
12	7	79	0.0%		11.3
13	0	0	0.0%		
Total	22,561	1,054,776	100.0%		

HILCC:
A Hierarchical Interface
to Library of Congress Classification

Stephen Paul Davis

SUMMARY. This paper describes the first phase of a project at Columbia University Libraries to create a "hierarchical interface to LC classification" (HILCC). The project's objective was to assess the potential of using the Library of Congress classification numbers as provided in standard catalog records to generate a structured, hierarchical menuing system for subject access to resources in the Libraries' electronic collections. The classification mapping table–jointly developed by the Libraries' systems, cataloging and reference staff–links each LC classification range with entry vocabulary in a three-level subject tree. Classification numbers and other metadata elements are extracted from catalog records in the Libraries' OPAC on a weekly basis, matched against the HILCC mapping table and then used to create browsable subject category menus to guide users to e-resource subject content. *[Article copies available for a fee from The Haworth Document Delivery Service: 1-800-HAWORTH. E-mail address: <getinfo@haworthpressinc.com> Website: <http://www.HaworthPress.com> © 2002 by The Haworth Press, Inc. All rights reserved.]*

Stephen Paul Davis is Coordinator, Columbia University Libraries Digital Library Initiative. He was Director of Library Systems, Columbia University and Analyst, Network Development & MARC Standards Office, Library of Congress.

The author would like to thank Bob Wolven and Patricia Renfro, both of Columbia Libraries, for useful and timely feedback on this paper.

This paper is a project report on the development of an operational prototype LCC-based subject interface to electronic resources at Columbia University Libraries.

[Haworth co-indexing entry note]: "HILCC: A Hierarchical Interface to Library of Congress Classification." Davis, Stephen Paul. Co-published simultaneously in *Journal of Internet Cataloging* (The Haworth Information Press, an imprint of The Haworth Press, Inc.) Vol. 5, No. 4, 2002, pp. 19-49; and: *High-Level Subject Access Tools and Techniques in Internet Cataloging* (ed: Judith R. Ahronheim) The Haworth Information Press, an imprint of The Haworth Press, Inc., 2002, pp. 19-49. Single or multiple copies of this article are available for a fee from The Haworth Document Delivery Service [1-800-HAWORTH, 9:00 a.m. - 5:00 p.m. (EST). E-mail address: getinfo@haworthpressinc.com].

19

KEYWORDS. Library of Congress classification, subject access, end-user interfaces, hierarchical subject menus, metadata, Columbia University Libraries

INTRODUCTION

Those who follow the progress of library-based information access and retrieval technologies will, if pressed, be obliged to admit that libraries and the automated system vendors that serve them have done little in the last decade to improve subject access to our print and, now, online collections. Much has of course been written and proposed in the library and information science literature about possible new strategies for access and retrieval, but few new approaches have actually been developed, tested and implemented in recent generations of library OPACs. Some would attribute this variously to: the marginal economics of library automation's niche marketplace; the timid approach vendors have taken to their feature enhancement processes; the enormous technical infrastructure changes libraries and vendors have had to absorb over the last ten years in order to stay even minimally current with new technologies; the aging systems of classification and subject analysis that continue to serve as our cataloging standards; the difficulty of innovating in OPACs when developers are constrained by the heavy hand of Z39.50 and fear the loss of interoperability with consortia and other cooperative systems; and the rise of the Web and the seemingly universal appeal of know-nothing, shot-in-the-dark keyword-Booleanism.

Even the traditional library strategy of creating and displaying "syndetic structures" of cross-references has foundered in the online environment. While most vendor-based library systems do now at least support the loading and integrated display of subject cross references in the OPAC, keeping them current and adequately customized to the local collection has proven to be too expensive and time consuming for many libraries. In institutions that have actually made this investment, the dismaying truth is that retrieval set displays can sometimes be overwhelming for users because of clumsy OPAC design and functionality. (A subject search for *political science* in virtually any large university's OPAC illustrates the problem nicely.) Research library users who bravely attempt subject searching are apt to encounter screen after screen of see-also references listed in alphabetical (i.e., conceptually random) disorder, followed by repetitive subject heading entries trailing endless subdivisions. Library patrons may be forgiven if they turn to the OPAC's keyword search function instead, or worse, flee to Google.

In this context it seems telling that, as soon as the Web became generally available, librarians almost instantly began coding and uploading informal subject-oriented lists of resources, whether of online databases, ejournals, or "Internet pathfinders." Perhaps unsurprisingly, these initiatives tended to come not from cataloging departments but from reference staff, selectors and collection development officers. The popularity of this approach was such that, as librarywebs grew and flowered, they would often sprout a number of different subject menus–overlapping, dissimilar–that were developed (and sometimes even maintained!) by different staff members or departments.

With the stunning growth of both commercial and noncommercial electronic resources and the commensurate need to collect and make them accessible to library users, the impossibility of 'scaling up' the creation and maintenance of manually created electronic resource lists has become increasingly apparent. In response to this, some libraries have already come to the conclusion that library cataloging–whether vendor-supplied, shared or original–may in fact be the best resource for meeting the challenge of providing flexible access to our burgeoning virtual libraries, and in a way that preserves conceptual integration with our enormous print and other non-electronic collections. The role of cataloging may not be as different in the near term future as some have predicted.

Recognizing the critical role of cataloging departments in allowing libraries to come to grips with issues of scale (as well as those of consistency and quality control) does not, of course, solve the problem of providing better subject access to our electronic collections. However, it does at least clarify the ground rules.

We can hope that the new generation of OPACs and their companion digital library platforms, such as Endeavor's ENCompass, will be able to deliver more flexible and effective user interfaces while also providing broad integration of access to our electronic and print collections. It is heartening to see that some commercial information vendors and aggregators, including OCLC and RLG, have in fact made great strides in recent years in enhancing their proprietary interfaces, improving functionality and design and providing users better guidance during the search process. These companies' deep pockets and commitment to product improvement have clearly benefited our patrons; at the same time they make the disparity between their systems and our OPACs even more glaring.

In response to these trends, a few libraries have begun to experiment with exporting cataloging data from their LMS's into newer, non-library database platforms and toolsets as a way to create more innovative Web-based access to electronic resources outside the context of the OPAC. Since these experiments are unhindered by the constraints of traditional library automated systems, they may yield interesting and useful new approaches to exploiting the subject and other metadata available in catalog records–approaches that may perhaps eventually find their way back into library OPACs.

This paper describes one such project, undertaken by Columbia University Libraries, to develop a system for the automatic generation of browsable, Web-based, subject-oriented presentations of our electronic resources by mapping LC classification numbers present in catalog records to a hierarchical subject vocabulary derived from the Library of Congress classification schedules.

BACKGROUND

In 1997, Columbia Libraries developed an SQL database and publishing system–called the Master Metadata File (MMF)–to support the automatic generation of Web pages listing reference databases, online reference texts and other electronic resources as they were added to our digital library collections. Each online reference resource was described in a brief "profile" record that was keyed into the MMF via a secure, Web-based CGI form. Normally this process–both the description and the keying–was done directly by the individual reference librarian or selector who had assumed overall responsibility for the title. This brief "profile" record was in fact quasi-AACR2 MARC-compatible metadata; but since public services staff members were creating the records we decided it was more politic to call these descriptions profiles rather than metadata records.[1]

As part of profile creation, the staff person was asked to assign one or more broad subject categories to the resource (see Table 1) as well as one or more standard resource type or genre designations such as "Abstracting & Indexing Service," "Ejournal" or "Full Text Resource." The profile also included additional description and annotation, including scope, keywords, search tips and related resources.

From this information we batch-published HTML listings by title, by subject category and by resource type. All links in these browsable listings, when clicked, took the user to an individual "about" screen–also batch generated from the MMF–that contained a description of the resource along with a stable, proxied (etc.) persistent URL link in the form of a "connect" button (see sample profile in Appendix A). HTML listings and about screens were regenerated on an ad hoc basis–usually several times a week–whenever any resource profile was added, changed or deleted.

This approach to producing browsable subject guides to our electronic reference resources worked well for a time, but inevitably began to show its weakness as the number of reference databases grew and as we began offering

TABLE 1. Columbia LibraryWeb Subject Categories, ca. 1997

Subject Categories	
Art, Architecture & Music	Law & Legislation
Business & Economics	Science & Technology
General & Interdisciplinary	Social Sciences
Humanities & History	

increasing numbers of electronic journals and texts to our users via the Web. We needed to move to a more scalable and less manual solution for access to electronic resources. We also needed a way to replace the informal subject categorization we had used so far with something more authoritative and comprehensive but also more specific and "granular."

In planning for ways to support an expanded Web-based presentation of our resources, we were aided by the Libraries' strategic commitment–which had evolved over the same period–to continue to provide standard cataloging for as many of our electronic resources as feasible, whether through shared, original or vendor-supplied records. What this meant in concrete terms was that we could anticipate relieving selectors and reference librarians of the need to create increasing numbers of "profile records," and instead extract MARC records for these same electronic resources from our LMS for loading into the Master Metadata File. Once loaded into the MMF, we would then be able to use standard programming and scripting tools (such as PERL, C++ and Java) to write batch or real-time interactive Web-based presentations for our users.

Among other benefits, this approach would give us access to the various subject-oriented elements present in these records–not only subject headings but also geographic area codes, contents notes, and LC classification numbers. We would still continue to provide a way for selectors and others to create metadata records directly in the MMF, but this would be limited to resources not considered appropriate for full cataloging (e.g., ephemeral Web sites or commercial services under evaluation) or those that had simply not yet gotten to the head of the cataloging queue (see Chart 1).

Our goal in expanding the MMF project in this way was *not* to develop anything like a fully functional end-user search and retrieval system that duplicated OPAC functionality. We did, however, see it as an opportunity to experiment more flexibly with the Web-based presentation of our electronic resources. One of our first priorities was to investigate ways we might exploit the subject content carried by LC Classification (LCC) numbers.

CHART 1. Workflow for Extracting E-Resource Records from the LMS for Loading into Columbia's Master Metadata File (MMF) Where They Are Accessible from Subject Category Menus via HILCC

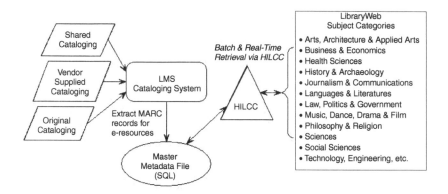

LC CLASSIFICATION[2] AS THE BASIS
FOR A HIERARCHICAL SUBJECT INTERFACE

We chose to focus on classification rather than subject headings because classification seemed better suited to our proximate goal, namely the creation of a browsable subject interface in which each electronic resource would appear under the one subject category that best reflected its content. We felt that this type of single-entry listing would provide a simple, easily navigable approach for many kinds of subject discovery, and would allow us to work toward replacing the various manually-created subject menus and listings in our libraryweb with an automatically generated subject interface. Moreover, we already knew from our experience with vendor-supplied OPACs the apparently insuperable difficulty of developing user-friendly search and retrieval interfaces for LC subject headings. Although we did in fact extract and load LC subject headings from catalog records into our MMF, our immediate purpose was to use them as enriched keywords for indexing by our Web-based search engines.

At about the same time that Columbia began working on this project, librarians at other institutions were evaluating the use of Dewey Decimal Classification (DDC) as a possible hierarchical system for access to Internet resources.[3] It appeared to some that DDC would lend itself to this type of subject browsing better than would LCC because of Dewey's intrinsically hierarchical design. Even if true, however, no one could possibly envision Columbia or other ARL libraries changing over (or back) to Dewey for all their cataloging, or even just the cataloging of electronic resources.

As we studied LCC, we saw–as Vizine-Goetz and others had observed–that underneath its enumerative top-level, there were potentially useful hierarchies at the second and subsequent levels as revealed by captions and the indispensable, conceptually significant indentations present in the schedules as a guide to classifiers. For example, the list of LC class number ranges shown in Table 2 actually reflects the useful hierarchy shown in Table 3.

On the other hand, many LCC hierarchies do seem outdated or at least less useful in the sense that naive users might have trouble inferring lower level subject categories from the upper levels of the hierarchy. For example see Tables 4, 5 and 6.

Beyond these types of hierarchical infelicities, LCC has well-known problems of terminology, consistency, balance and what might be called "world view," many of which have been discussed in the literature, at conferences, and doubtless internally at the Library of Congress. Some examples are on the following page.

TABLE 2. Selected Class Ranges from LCC's Philosophy, Psychology, Religion Schedule

B
BM
BM 495-532
BM 497-509
BM 497-497.8
BM 498-498.8
BM 499-504.7
BM 507-507.5
BM 508-508.5

TABLE 3. LCC Captions Corresponding to the Class Ranges in Table 2

Philosophy, Psychology, Religion.			
	Judaism		
		Sources of Jewish Religion, Rabbinical Literature	
			Talmudic Literature
			Mishnah
			Palestinian Talmud
			Babylonian Talmud
			Baraita
			Tosefta

Class **D, "History: General and Old World."** A single class schedule, D, is allocated for the history of all regions of the world, apart from America; two entire schedules, E and F, are dedicated to America and the United States. This certainly places the cultural and conceptual framework of those who created in LCC in stark relief. At the very least the term "Old World" must now be seen as outdated and culturally biased toward the Western Europeans and Americans who invented the New World/Old World dichotomy.

Class **E, "History: America."** Toward the beginning of schedule E a hierarchy appears as shown in Table 7.

This is probably not a hierarchy that one would want to present to users directly.

Class **B, "Philosophy. Psychology. Religion."** Under Class B no fewer than one-third of the second level terms are given over to Christianity, including one for "The Bible," which is not even identified as the Christian Bible. And while there may once have been excellent reasons for sandwiching Psychology

TABLE 4. Excerpt from LCC's Geography, Anthropology, Recreation Schedule

G		Geography. Anthropology. Recreation.	
GT		Manners and Customs (General)	
GT 485			Churches and church going
			etc.

TABLE 5. Excerpt from LCC's Geography, Anthropology, Recreation Schedule

C		Auxiliary Sciences of History	
CB		History of Civilization	
CB 156			Terrestrial evidence of interplanetary voyages
CB 158-161			Forecasts of future progress
			etc.

TABLE 6. Excerpt from LCC's Technology Schedule

T		Technology	
TX		Home Economics	
TX 341-641			Nutrition. Food and Food Supply
TX 950-953			Taverns, Barrooms, Saloons
TX 1100-1105			Mobile Home Living
			etc.

between Speculative Philosophy and Aesthetics, the intercultural rudeness of including Theosophy–an eccentric American cult dating from the late 19th century–in the same caption as Islam might at the least bewilder our patrons (see Table 8).

So in short, our working group realized quickly that we would not be able to use LCC "out of the box," as it were, to create a browsable user interface. Rather than being discouraged, however, this recognition was in fact somewhat freeing. It meant that, while we would indeed be able to exploit the LC class numbers themselves, the notion of using LCC's language and structure more generally was so clearly out of the question that to proceed at all we would have to develop different entry vocabulary and create alternate hierarchies.

Since we planned to build this presentation hierarchy outside the cataloging system and then process catalog records with LC class numbers against this mapping schema, we would have the freedom to experiment with different presentation strategies–adding to, changing and rearranging as we needed. We could easily, for example, reformulate the Philosophy and Religion presentation (see Table 9).

Or we might take a more taxonomic approach, where for example Christianity and Islam branch from Judaism; or a highly enumerative, alphabetic approach giving equal time to each religion and sect. This realization helped inspire us to continue our planning.

PROJECT PLANNING

A working group with members from our Bibliographic Control Department and Library Systems Office was convened in 1997 to begin working on a "Hierarchical Interface to LC Classification" (HILCC).[4]

Our informal objectives for this Phase I pilot project were to:

- Develop a preliminary classification map and subject category hierarchy for each LCC schedule;
- Invite reference staff, selectors and other subject specialists to assist with reviewing and revising the various subject areas of this preliminary HILCC map;
- Extract a test dataset of catalog records for ejournals from our LMS;
- Convert and load test record sets into our Master Metadata File;
- Specify and program one or more Web presentation options;
- Review results and revise;
- Put HILCC into production as an operational prototype;
- Continue to gather staff and user feedback; continue to make corrections, updates and enhancements;

- Operationalize the weekly extraction and loading of catalog records for ejournals from the LMS into the MMF and the regeneration of the browsable HILCC interface to our eresources;
- Expand the dataset to include other e-resources beyond ejournals.

Almost immediately, however, as we faced the daunting task of coming to grips with the full range of LC class schedules, we felt the need to place certain kinds of limits on our work in order to make the project manageable within a reasonable timeframe, especially since the work was being done by staff fully engaged with other primary responsibilities. Some helpful working assumptions that evolved were:

1. In the current phase of the project, we would create no more than three hierarchical levels at any point in the schema, even if it might be desirable to go to a fourth or fifth;
2. Our main goal was to provide an effective end-user interface to LC Classification numbers, not to spend time rearranging LCC itself;
3. It was not our goal to create a universal subject interface, only one that would serve the Columbia community;
4. We should not attempt to duplicate functionality already available in the current generation of OPACs;
5. We should proceed by creating operational prototypes that could be assessed and improved over time;
6. HILCC should be seen as only one of several possible ways for our users to search for resources by subject; it did not need to solve all or even most subject retrieval problems.
7. We more or less managed to keep these guidelines in view as we proceeded.

WORKING DESIGN PRINCIPLES AND CONSIDERATIONS

Perhaps the best way to describe the process of developing HILCC is to review the design principles that evolved as we worked. Some of these ideas had

TABLE 7. Excerpt from LCC's History: America

E	History: America		
E 151-889	United States		
E 184-185.98		Elements in the population	
E 184.5-185.98			Afro-Americans
E 186-199		Colonial history (1607-1775)	
		etc.	

TABLE 8. Excerpt from LCC's Philosophy, Psychology, Religion Schedule

B	Philosophy. Psychology. Religion
BC	Logic
BD	Speculative Philosophy
BF	Psychology
BH	Aesthetics
BJ	Ethics
BL	Religions. Theology. Rationalism
BM	Judaism
BP	Islam. Bahaism. Theosophy, etc.
BQ	Buddhism
BR	Christianity
BS	The Bible
BT	Doctrinal Theology
BV	Practical Theology
BX	Christian Denominations

been articulated at the outset of the project, but for the most part they grew out of questions and issues raised by the ongoing work itself. The discussion below is organized around individual design and development decisions made during the course of the project.

The first level display should include no more than twelve (or so) categories. We felt that in many respects, HILCC's first-level menu organization was the most important design challenge. People's ability to find relevant resources without repetitive backtracking depended directly on their ability to choose the correct top branch of the hierarchy. The first menu needed to be clear enough for undergraduates but not too simplistic for faculty and graduate students searching outside their usual domain. It needed to be balanced in terms of disciplines and subject areas, but also reflect Columbia's actual online collections and academic curriculum.

We began with an informal survey of other university and commercial sites that already provided general-purpose browsable listings by subject category to see what could be learned. With regard to the specific question of how extensive a top-level listing might be, we did come away with some impressionistic results. The following list shows the number of first-level subject category menu items we found in a typical selection of sites. (The following numbers reflect the sites' first level menus as of December 2001, but the counts are virtually the same as those we obtained in 1997. Shopping and other non-substantive links have been omitted.)

MSN Search - 10	BUBL (Strathclyde University) - 10
Hotbot - 14	NC State University - 12
Yahoo! - 14	LC American Memory - 13
Excite - 15	WWW Virtual Library - 14
Netscape Search - 16	

TABLE 9. Philosophy & Religion Section of HILCC

HILCC Presentation Hierarchy:			Corresponding LC class number ranges:
Philosophy & Religion			
	Philosophy		
		Philosophy (General)	B 1-5802
		Aesthetics	BH 1-301
		Ethics	BJ 1-2195
		Logic	BC 1-199
		Speculative	BD 1-701
	Religion		
		Religion (General)	BL 1-290, BL 350-632.5
		African Religions	BL 2390-2490
		Ancient Near Eastern Religions	BL 1600-1695
		Bahaism	BP 300-395
		Buddhism	BQ 1-9800
		Christianity	BR 1-1725, BS 1-2970, BT 10-1480, BV 1-5099, BX 1-9999
		European, pre-Christian Religions	BL 689-980
		Hinduism	BL 1100-1299
		Islam	BP 1-253
		Jainism	BL 1300-1380
		Judaism	BM 1-990
		Mythology, Comparative	BL 300-325
		North & South American Religions	BL 2500-2592, E98.R3
		Oceania Religions	BL 2600-2630
		Rationalism, Atheism, Secularism	BL 2700-2790
		Other religions	BL 660-687, etc.

Based on these results, we felt comfortable proceeding with a menu of between ten and fifteen top-level items, finally settling on twelve as our working target. The number and content of those categories did evolve considerably during the course of the pilot, however, as we received feedback from different groups of library staff. Still we did ultimately end up with a twelve-item main menu.

We also decided early on that we would attempt to follow several principles that were (and still are) considered to be good design practice for Web page usability, namely (a) that Web pages should ideally take up less than one "screen," without scrolling, on a standard 600 x 800 pixel display; and (b) that the user should ideally be able to reach any desired item in no more than three mouse clicks from the main page; and (c) that generous use of white space as a design element is pleasing to the eye and helps avoids visual clutter. As many others have found before us, applying all three of these principles simultaneously is often virtually impossible, particularly when working in a complex and deeply hierarchical domain. We can say with a high degree of confidence that we were not entirely successful in achieving these design goals.

Since our starting point was LC Classification with its twenty-one separate schedules, we were immediately obliged to combine and merge HILCC categories in ways that we hoped would yield a list with straightforward terminology; also one that would allow users to easily discern the likely conceptual boundaries of each category. The results of this exercise in redividing the world of knowledge may be seen in Table 10.

In most cases our selection of categories seemed reasonable and sensible for the Columbia environment; a few however were recognized from the outset to be less effective and were flagged for early review during Phase II in conjunction with user assessment studies. Terms that might have been useful as unifying top-level categories were sometimes rejected because of their perceived unfamiliarity to users and replaced with an enumeration of second level terms. For example, we initially selected the term "Applied Sciences" to cover technology, engineering and computer science; but informal student feedback made it clear that the term "applied sciences" is not only *not* generally used, it was considered utterly opaque by our users. So we fell back here as elsewhere on a caption that enumerated rather than summarized the major subareas, i.e., Technology, Engineering & Applied Sciences. Likewise, the more generally used rubric "Performing Arts" that is often used to describe music, dance, drama and film was deemed confusing at Columbia, chiefly because these areas are for the most part not actually taught as performing arts here. (Columbia's major areas of music in-

TABLE 10. HILCC Level 1 Categories (ca. 2001)

Main Subject Categories	
Arts, Architecture & Applied Arts	Law, Politics & Government
Business & Economics	Music, Dance, Drama & Film
Health Sciences	Philosophy & Religion
History & Archaeology	Sciences
Journalism & Communications	Social Sciences
Languages & Literatures	Technology, Engineering & Applied Sciences

struction and research, for example are ethnomusicology, historical musicology, music theory and composition.)

The process of settling on top-level categories proceeded iteratively over time and left those of us involved with a new appreciation of the inherent difficulties in trying to organize knowledge according to a generalized schema. The categories chosen for the top level of course reflect biases of individuals in the working group and other library staff who contributed to the effort. They also reflected explicit and implicit collection development policies, which in turn reflect the curriculum and research priorities of Columbia University.

For example, subject areas relating to many of Columbia's major professional schools, which are served by the central library system–such as Architecture, Business, Engineering and Journalism–figure prominently in top-level categories. Columbia's Health Sciences Library and Law Library are relatively independent from the main library and did not participate during Phase I in the development of HILCC, with the consequence that those sections of HILCC are less well developed and adhere more closely to the LC Classification outline than they might have otherwise.

Some library staff have suggested that our top-level organization is weighted disproportionally toward the humanities; this is something we will revisit during Phase II. How we would in fact be able to determine what an equitable balance of top-level categories might be remains, at this point, mysterious.

Additional biases in HILCC can easily be recognized by a simple list of other institutions' top-level categories that do *not* appear in Columbia's, e.g., Agriculture, Bibliography/Library Science, Computer Science, Education, Generalities, Geography, Military Science, Naval Science, Recreation, Sports, Reference, Veterinary Medicine.

HILCC's overall hierarchy should be no more than three levels deep. It was clear from the beginning that some sections of LCC lent themselves, in theory, to four, five and six-level hierarchies. For pragmatic reasons and usability considerations we decided to limit ourselves to three hierarchical levels in Phase I of the project.

In some cases we did this by skipping over hierarchies implied in LCC, e.g., for class ranges BL1000-BL1299 we used the simpler:

Philosophy & Religion – Religion – Hinduism
   rather than LCC's:
Philosophy & Religion – Religion – History & Principles of Religion – Asian, Oriental
Hinduism

Some extensions of HILCC to a fourth or even fifth level do seem inevitable, however, and are already in the planning stages for Phase II. We will likely interpose levels of geographical hierarchy in the history portion of HILCC, and a few other places. For example, for the class ranges DD1-9999, we anticipate using:

History & Archaeology – Regions & Countries – Europe – History of Germany

rather than LCC's:

History: General & Old World – History of Germany

It is clear that we were not able to meet our ideal page length target of "one screen" when enumerating at the second and third levels of the HILCC hierarchy. In Phase 2 we will be aiming for more compact and flexible presentation techniques, such as JavaScript rollovers, expandable folder hierarchies, etc.

The degree of granularity should be relative to the actual resources available. In determining how enumerative and specific to be at any given hierarchical level, we decided to be guided by the current depth of our electronic holdings in a specific area. We recognized in doing this that we would need to revisit and revise portions of the hierarchy periodically as larger numbers of resources were added to our digital collections, but felt this would be an important and useful process in any event.

As a separate but related issue, if at a given level of HILCC–at whatever the degree of granularity implemented–there were in fact categories for which there were no electronic resources yet available, we decided to retain and display these unused categories as "grayed out" menu items on the assumption that this would aid users in recognizing that they had indeed navigated to the correct place in the hierarchy, but that there were no "hits" for their topic, e.g.,

Arts, Architecture & Applied Arts

Visual Arts (all)

General

Decorative Arts
Drawing, Design & Illustration

Painting

Photography
Print Media

Sculpture

Subject categories should be built from LCC but not constrained by it. It was apparent to us that in some cases the existing structure of the LCC tables would indeed provide the basis for a reasonably coherent user presentation. In others, however, we would either need to ignore the explicit or implicit hierarchies present in LCC schedules or else reorganize them in order to create a balanced and usable interface. For example, LCC's schedule G, "Geography, Anthropology and Recreation" did not seem a useful category grouping for end user presentation & navigation, and the G class ranges were variously reassigned to the sciences and social sciences.

A specific LC class range should map to only a single location in the HILCC structure. We initially attempted to map some classification ranges into more than one HILCC category, but realized that this would add substantially to the maintenance overhead and the programming requirements of the project. We also recognized that the issues raised by this kind of double mapping were really just the beginning of a larger analysis and planning effort that would be needed to address the challenge of presenting interdisciplinary resources to users (see also section g following).

An area in which we were especially tempted to map a single class range to more than one HILCC location was that of Psychology (BF). LCC of course embeds Psychology in the middle of Philosophy, quite near Religion and "The Occult Sciences." Although it is perhaps possible that this was once the ideal location for it, modern psychology has subdisciplines that fall variously into the social sciences, the sciences and the health sciences. In HILCC we finally decided to position Psychology under the Social Sciences; but because Columbia's Psychology Library has traditionally been designated a science library, we also manually created a duplicate link in our user interface from within the Science hierarchy.

HILCC processing and output should accommodate multiple LC class numbers appearing in a single bibliographic record. Although standard catalog records rarely include more than one classification number for the purposes of capturing different subject aspects of the same work, we anticipated this occurring on an ad hoc basis at Columbia as selectors and reference staff requested that important electronic resources be available under different subject trees.

In fact, we had previously established a precedent for multiple subject category assignments in our initial Master Metadata File implementation where, when selectors created profiles for individual reference databases, they were able to select up to three subject category areas under which the title would be listed. In part this was a reflection of the broad scope of some of the large databases; *Annual Reviews*, for example, contains extensive resources in sciences, health sciences and social sciences.

The categorization and presentation of interdisciplinary resources should be addressed separately from the main HILCC effort. Reference staff and selectors consulting on the HILCC project emphasized to the working group the importance of also providing better and more comprehensive displays of interdisciplinary resources than could be derived from a generalized mapping of LCC numbers. This need had so far been addressed at Columbia by the manual compilation and maintenance of specialized guides or "Internet pathfinders" on interdisciplinary topics such as women's studies, African-American studies, Middle East studies, etc.

We sketched out a possible method of addressing this need through the creation of separate, customized versions of HILCC for the different interdisciplinary areas. These selective classification maps could be used–perhaps in conjunction with other elements found in the catalog record–to extract, filter and display more targeted presentations of subsets of our electronic resources.

It did seem, however, that this would depart somewhat from the project's primary objective of creating a generalized subject interface to all our electronic resources. It also seemed clear that, even more than with the generalized HILCC interface, these interdisciplinary "mini-HILCCs" would require a higher level of ongoing maintenance since they would inevitably need to reflect Columbia's evolving academic organization and curriculum, the specific collection development policies of our different collections and departmental libraries, not to speak of the particular strengths and interests of the library staff members responsible for particular subject areas. The working group agreed that this was an important area that we would nonetheless need to postpone to Phase II.

The user interface must include composite, summary lists at the first and second levels of each hierarchy. Once a user manages to locate the "correct" top-level category, it still may be unclear how far down a hierarchy to drill to find a specific narrower concept; or in some cases the user may just want a complete view of the subject at a second or third level. We thus assumed more or less from the beginning that we would need also to provide combined listings at the first and second levels. In this way, users who give up before finding their way down to the lowest level of the hierarchy still have a chance of finding relevant resources.

For example, if a user is trying to locate electronic resources dealing with European public policy, and has chosen the correct top-level menu category, she or he might navigate down the following hierarchy:

Law, Politics & Government – Political Institutions & Public Administration – Europe

But if the user gives up before drilling all the way down to "Europe," all relevant resources on that specific topic would still also be findable under: *Law, Politics & Government (all)* as well as *Political Institutions & Public Administration (all)*. These higher-level artificial composite listings might of course be lengthy, but the desired resources would at least be present and findable through determined browsing or judicious use of the browser-based find command.

When feasible, terminology used at the lowest level of the hierarchy should be meaningful and unambiguous when displayed independently. In order to have the capability of flexibly displaying and repurposing HILCC output in different contexts, we came to the conclusion that the terminology at the lowest level of the hierarchy, at least, should be constructed such that it could be displayed independently of the hierarchy. For example, we would use:

Social Sciences / Education / History of Education

rather than the elliptical:

Social Sciences / Education / History

The term "History" used alone in a list or display would obviously be ambiguous or misleading, whereas "History of Education" could stand on its own.

One of the immediate reasons for this decision was that we wanted to use these lowest-level HILCC terms to produce a separate topical subject index to be used in conjunction with HILCC, since it was clear that a topical list would be easier for some kinds of resource discovery–particularly when the terminology and conceptual organization of the subject was straightforward (see Table 11).

Planning for and implementing a successful "inheritance" schema for displays of hierarchical data is complex and not something we felt we could necessarily take for granted in the design of HILCC. We recognized, though, that the decision to create stand-alone terminology would itself lead to more questions about how we should formulate these lowest-level terms when there was no obvious equivalent in common usage, or when phrases might perhaps be inverted to help cluster-like content in an alphabetic listing, e.g.,

History of Education

[. . .]

Theory and Practice of Education

or:

Education, History of

Education, Theory and Practice of

TABLE 11. HILCC-Derived Topical List

Topical list:	From corresponding HILCC category:
[...]	
Biochemistry	Sciences -- Chemistry -- Biochemistry
Bioengineering	Technology, Engineering & Applied Sciences -- Engineering -- Bioengineering
Biography	History & Archaeology - Biography
Biology	Sciences -- Biology (all)
Biomedical Engineering	Health Sciences -- Biomedical Engineering
Biophysics	Sciences -- Biology -- Biophysics
Book Studies & Arts	Social Sciences -- Education & Scholarship -- Book Studies & Arts
Botany	Sciences -- Botany
Business	Business & Economics -- Business (all)
[...]	

In Phase 2 we will compare terminology at the lowest level against LCSH to see if corresponding terms can be borrowed for use in HILCC. We may find that we need to store two versions of the lowest-level term, one suitable for display in a hierarchy, and one capable of standing alone without ambiguity.

IMPLEMENTATION ISSUES–SCIENCES

Many of the types of issues that arose during HILCC development can be seen in miniature in the Sciences hierarchy, which was one of the first to be prototyped. A partial summary of decisions made in constructing this section follows along with brief characterizations of the discussions that led to them. (Refer to Appendix B, the Science portion of HILCC with a listing of corresponding LCC ranges; and Appendix C showing selections from LC Class Schedule Q with corresponding LCC captions.)

Terminology Changes

LCC's Level 2 caption "Astronomy" (QB) was expanded to "Astronomy & Astrophysics," essentially elevating a third level LCC category to the second level of HILCC. This change was made to reflect the combined orientation of

Columbia's academic and research programs in these fields and the Libraries' corresponding collecting policies in these areas.

LCC's Level 2 caption "Neurophysiology and Neuropsychology" was broadened to "Neurosciences," which was felt to be a more useful summary of the scope of material collected and classed in LCC's QP 351-495 range.

New Intermediate Groupings

We created an intermediate hierarchy under Sciences for "Earth and Environmental Sciences," into which we moved Ecology (QH 540-549.5) Environmental Sciences (GE 1-350), Forestry (SD), Geology (QE), Human Ecology & Anthropogeography (GF), Meteorology & Climatology (QC 851-999), Natural History (QH 1-278.5), Oceanography (GC) and Physical Geography (GB).

This collocation was undertaken in part because "Earth and Environmental Sciences" has become an important focus at Columbia both in terms of academic organization and curriculum focus. Since this is clearly a cross-disciplinary grouping, however, we may want to revisit this in Phase II when we have the capability of creating separate, customized interdisciplinary mini-HILCCs.

We combined LC's "Human Anatomy" and "Physiology" sections into a single HILCC category "Human Anatomy & Physiology" at the second level, moving Anatomy, Neuroscience and Physiology to separate subcategories at the third level.

A new Level 3 category "Mathematical Statistics" was created from the LCC range QA 273-280.999, which includes LCC captions "Probability" and "Mathematical Analysis." This was done so that different areas of the classification schedule relating to statistics of various kinds could be brought together with one another, e.g., in generating an overall guide to electronic resources in statistics and quantitative methods. Again, it seems likely that the interdisciplinary "mini-HILCCs" envisioned as part of Phase II might lead us to reconsider whether to retain this type of artificial category in the master version of HILCC.

Rearrangement of Hierarchies

Biochemistry and Radiation Chemistry are at the fourth level of LCC captions; in HILCC they were moved up to the third level directly under Chemistry for HILCC, so that they did not disappear from HILCC's three-level hierarchy.

Microbiology (QR) was moved from Level 2 in LCC to Level 3, under Biology, in HILCC.

Merging of Schedules

Much of LCC's Class S (Agriculture) was folded into HILCC's "Sciences" hierarchy under two new intermediate groupings "Animal Sciences" and "Plant Sciences." (Note, however, that Botany and Zoology were retained in this Phase as Level 2 categories, chiefly because of their traditional places in science collections and curricula. As ongoing review of HILCC proceeds, such decisions may well be reconsidered.)

The decision to merge Class S into the HILCC "Sciences" hierarchy was made largely because materials relating to Agriculture and related fields are not collected in any significant way by Columbia Libraries nor has Agriculture figured significantly in Columbia's curriculum. The rationale for this decision illustrates concretely why different institutions might organize HILCC-like subject hierarchies quite differently. Still, Columbia does in fact have printed and electronic resources relating to Agriculture, notably those that we receive as part of the Federal Depository Library Program as well as those that appear unbidden in ejournal aggregator packages such as ProQuest. Moreover, Columbia's newer curricular and research emphases on Earth and Environment Sciences include the effects of global climate change on agriculture, so again, HILCC may need to evolve further in this area as elsewhere.

NEXT STEPS

Assessment. Although no decisions have been made about a Phase II for the HILCC project, some kind of user assessment will be needed, whether separately or as part of other planned LibraryWeb usability testing. We will shortly have some 5,000 records available in our Master Metadata File–for databases, ejournals, e-texts, etc.–that we will make accessible through the HILCC interface; this should provide the critical mass needed for a targeted user assessment. Even so, the effectiveness of HILCC may be difficult to test. The success of an individual user's interaction with HILCC may depend on a number of factors, including:

1. the appropriateness and relevance of the classification numbers originally assigned to our electronic resources;
2. the aptness and recognizablity of the specific HILCC terms Columbia selected for that specific LC class range;
3. the transparency of the hierarchy in which Columbia positioned the specific HILCC subject term;
4. the usability of the specific Web design and navigation functionality which we build into and around HILCC;

5. the placement and description of HILCC in Columbia's LibraryWeb relative to other types of subject searching, and the way the user's expectations are shaped in advance of using it;
6. the length, organization and transparency of the list of resources retrieved after navigating a hierarchy;
7. the actual presence or absence of specific electronic resources in our collection of interest to the user.

There are also the many user-based variables–such as the user's purpose in searching, past online experience, other systems used, etc.–which we already know how to capture and record if not always how to interpret.

But what may be impossible to design a test for is the overall value of browsing electronic resources by subject categories in the first place. In the brick and mortar world, the value of organizing print collections for browsing has always been difficult to quantify; but libraries have nonetheless spent many millions of dollars over the decades in order to create and maintain them. When questions have arisen about the cost-benefit of classified collections, the arguments in favor cited convenience, serendipity, benefit to the non-specialist and such. Do these values translate to the online environment?

There are some practical considerations bearing on the effectiveness of a HILCC-like approach that are difficult to anticipate, most notably the issue of scalability: what may seem useful and manageable against a list of 5,000 electronic titles may look quite different when the list has grown to 50,000 or more.

Another even more basic concern may be the actual availability in source records of the LC Class numbers themselves. As large libraries begin increasingly to purchase catalog records, for ejournals, etc., from aggregators or specialized service bureaus, they also become reliant on those companies to obtain or themselves assign relevant LC Classification numbers. At present this is not always the case, and the cost of adding class numbers to thousands of purchased catalog records would be prohibitive for most institutions.

Phase II. Assuming Columbia's HILCC project proceeds to a Phase II, the following improvements and extensions to HILCC would be in queue for consideration.

Subject Specialist Review. Continued review of the various subject areas of HILCC by library subject specialists and/or faculty as we add more electronic resources to our digital collections.

Interdisciplinary Mini-HILCCs. Perhaps the most urgent finding from our Phase I operational test was the importance of extending the HILCC project to include customized presentations of interdisciplinary resources. In research universities especially, new interdisciplinary institutes, centers and projects are created (and sometimes disappear) quickly. Library staff need a simple but flexible way of creating targeted listings of resources by computer-assisted means, both on-demand and for public presentation of our digital library collections.

Review of HILCC Terms Against LC Subject Headings. Work has already been done at the Library of Congress and elsewhere to begin to correlate LC subject headings with relevant portions of the LC Class Schedules. Improving HILCC terminology by use of LC subject headings may benefit users and provide the Library more options for using HILCC as a basis for newer kinds of subject-oriented access.

Resource Presentation Using Additional Metadata Elements. We recognize the importance of using other metadata elements in combination with HILCC for browsing and searching for materials; these additional elements or aspects include: genre or format (e.g., ejournals, e-texts), geographic content (e.g., on-line resources from or about sub-Saharan Africa), "reference-ness" (e.g., key online databases in Public Affairs).

New Interactive Modes of User Discovery. The availability of rich metadata extracted from catalog records and made available via a robust SQL/Web retrieval framework potentially provides a powerful toolset to experiment operationally with more "intelligent" forms of user interaction such as:

- basic research dialogues with users, allowing search strategies to be refined, expanded, limited, etc.
- content mapping and visual navigation, allowing users to see the depth of content in certain collection areas, the relationships between content clusters
- interactive query optimization with user-assigned relevancy weighting
- creation of a non-specialist cataloging interface for faculty or other researchers to prepare metadata and integrate their resources into Columbia's digital collections using HILCC categories.

CONCLUSIONS

Hierarchical, Term-Mediated Classification Systems. Judging solely by the continuing presence and even expansion of hierarchical subject-oriented menus in the major commercial search portals, one would have to come to the conclusion that this type of interface is considered an essential access tool by firms that have a more-than-altruistic interest in helping people locate resources. Taking Yahoo! as perhaps the most notable example: not only does its home page prominently feature its proprietary classification hierarchy, Yahoo!

also does something that we in Libraries should have been doing long since. In response to a keyword search, rather than presenting the user with a list of "raw" hits, there's a simple but far more intelligent response:

- Category Matches

  o   [hits]

- Sponsored Links

  o   [hits]

- Web Site Matches

  o   [hits]

"Category Matches," promoted to the top of the result screen, are in fact links back into Yahoo!'s classified hierarchy. The OPAC search for "political science" proposed in the introduction of this paper is, by contrast, a relief and a pleasure to execute in Yahoo! if one navigates via the category matches. Of course–this being Yahoo!–once you retrieve the actual resources at the bottom of a selected hierarchy, the content may well be outdated, eccentric or wrong; but the functionality and interface design that gets you there puts library OPACs to shame.

How enormously helpful it might be for patrons if our OPAC keyword searches returned:

Subject Category Matches (browse by subject category)
  [hits]

Subject Heading Matches (narrow your search)
  [hits

Reference Matches (key resources)
  [hits]

Individual Title Matches
  [hits]

Lest one think that Yahoo!–famously staffed by bona fide librarians–is an outlier, Google too has a powerful, if somewhat less developed, classification system that, again, displays at the top of the raw results screen the relevant category hits from their classification system. Ditto MSN, Overture (formerly GoTo), Galaxy, Oingo/Applied Semantics (an excellent search system!), Teoma and Webcrawler. In many respects creating a classification system simplifies the work of portal system designers and gives them a powerful and manageable conceptual toolsct. (If instead they had somehow been persuaded to wrestle with our tens of thousands of LC subject headings and their tangle of references, they too might have given up in despair.)

By contrast, keyword and subject heading searches by themselves more often than not yield ambiguous results. A keyword searcher is always, effectively, a beginner; one must proceed each time by trial and error, usually without knowing entirely whether one has succeeded or failed. The results are often mysterious; it is unclear whether one has selected the best keywords (or subjects or references). In the usual case, the searcher may indeed retrieve some results but never really know whether these are–even objectively speaking–all, some or any of the results most relevant to the topic at hand in the system being searched.

Hierarchical classification systems, if mediated by entry vocabulary, explanations and a well-designed navigation framework, can offer the benefit of more easily guiding users through a process of recognition, filtering and decision making than do other subject retrieval methods. Visible hierarchies can help show users how the information provider has actually organized the resources conceptually. Armed with this understanding, the user may learn more quickly and in real time how to find resources of interest than with other methods.

Hierarchical Interfaces to Library of Congress Classification. During the course of this project we were sometimes struck by the utter hubris that would be needed today to undertake from scratch the creation, implementation and maintenance of a generalized classification system that was: accurate and socially sensitive; learned but not esoteric; consistent, proportional and conceptually balanced; suitable for students and scholars (and citizens and members of Congress) . . . and capable of remaining all these things for more than a few months or years. But while we wait for a new Charles Ammi Cutter–and a new Herbert Putnam with the vision to push this forward–what we in research libraries have to work with for now is LC Classification.

The HILCC project at Columbia is still at a relatively early stage, particularly in terms of the possible functional integration of HILCC with keyword or guided searching methods. The process of LCC mapping and terminology selection that is at the heart of HILCC has been time-consuming, complex and is still not complete, accurate or authoritative even on its own terms. Still, I believe we have succeeded admirably in our original objective, namely to provide within our LibraryWeb an integrated, browsable subject-oriented presentation of our electronic resources.

In a broader context, it may well be feasible for others to take advantage of what we have learned from the HILCC project in at least two areas. Those institutions or consortia with the resources to build services outside their OPACs might consider taking Columbia's LCC map and reworking it for their own environments and collections.[5] If others do indeed take this on, Columbia would be grateful for feedback and suggestions and more than willing to answer questions or review proposals.

On the other hand, after having spent a great deal of time attempting to come to grips with the many small and large mysteries of LCC, such as how and why "Hyperborean, Indian, and Artificial Languages" are clustered together in PM 1-9021 and whether there's a more scholarly and/or socially sensitive way to present them to our users, we might be forgiven for wondering if the Library of Congress might after all deserve to take on the role of developing and managing an authoritative, general-purpose version of HILCC. In this imaginary scenario, LC would act as coordinator and co-developer as it does with various other standards, and ultimately act as distributor for the schema and its updates as it does with other cataloging-related products. Ideally HILCC would also be used directly within LC's own information systems as an interface to its own electronic resources. This would help ensure that HILCC was grounded in actual eresource collections and end-user interactions and, perhaps, given some priority for thoughtful development and refinement. Those of us on the outside who subscribed to the product service would be free to tweak it for their local environments and, of course, kibitz.

It might be, though, that LC has even more to gain from such an approach. It is conceivable that a tool such as HILCC could provide a platform with which to model revisions or perhaps even a complete overhaul of LCC. (Surely someday we deserve an LCC21 or at least an LCC22!) A key benefit would be that LC (and other) catalogers would be able to continue using "old LCC" indefinitely while schedules were being modeled, discussed, revised, discarded and reworked. HILCC itself provides the "switching system" or crosswalk between the old and new. One can imagine many different development and implementation possibilities worth exploring.

In conclusion, then, as libraries collectively hunker down and face the prospect of trying to manage and provide access to an exploding number of electronic resources with, as we are warned, fewer and fewer staff resources; and as we try to distill for ourselves and our parent institutions what it is exactly that we have to offer the information world apart from offsite storage facilities and eresource license management; let me suggest that classification, used alone and in combination with keyword searching, appears to be well suited to resource discovery within a broad heterogeneous information environment and, if developed intelligently, could be a key tool in our evolving knowledge-based environment.

NOTES

1. Examples of the various components of this project may be found in Carol Mandel's "Manifestations of Cataloging in the Era of Metadata" [outline & slides], ALCTS/LITA Institute on Managing Metadata for the Digital Library, May 4-5, 1998, available online at: *<http://www.columbia.edu/cgi-bin/cul/resolve?cul.1BILCW>*.

2. For the uninitiated, LC Classification was developed at the Library of Congress beginning about 1899 and was not, contrary to rumor, based on the organization of Thomas Jefferson's library but rather on Charles Ammi Cutter's *Expansive Classification* (1891). A convenient outline of LCC is available online at: *<http://www.loc.gov/catdir/cpso/lcco/lcco.html>*. LC Classification numbers constitute the first portion of LC call numbers and are carried in MARC21 field 050, subfield a.

3. See, for example, Vizine-Goetz, D. "Using Library Classification Schemes for Internet Resources." In: *Proceedings of the OCLC Internet Cataloging Colloquium*, San Antonio, Texas, January 19, 1996. URL *<http://staff.oclc.org/~vizine/Intercat/vizine-goetz.htm>*.

4. The working group currently includes the following Columbia Libraries staff members: Rick J. Block, Stephen P. Davis, Kate Harcourt, Sarah H. Witte, Robert A. Wolven; in addition, former staff member Jeffrey Sowder contributed to the development of HILCC.

5. Columbia's HILCC documentation and current version of the LCC map is available at: *http://www.columbia.edu/cgi-bin/cul/resolve?cul.1BQN3R*.

APPENDIX A. Sample Electronic Resource Profile (1997)

The metadata record for "Art Index" included the information displayed in the About profile below, and was designated by the library selector to appear under two subject categories from an early prototype of HILCC, namely:
    Art, Architecture & Music
    and
    General & Interdiscipinary

***About:* Art Index**

Coverage

Subject Description:	2 part index to art periodicals in Western European languages: Part 1 (420 journal titles): from 1929-1984; Part 2(376 journal titles): from 1984 to the present. Beginning in 1994, abstracts are included. After the late 1990s, some full text is available (lacking illustrations).
Subject Keywords:	archaeology, architecture, art history, city planning, computer graphics, crafts, film, folk art, graphic arts, industrial design, interior design, landscape architecture, museology, painting, photography, sculpture, television, textiles, video.
Years:	Part 1: 1929-1984; Part 2: September 1984 to the present with abstracts starting in Spring 1994, and some full text beginning in the late 1990s.
Content Type:	Citations, abstracts, and some full text.
Updated:	Part 2: updated monthly.
Regions:	U.S. and international.
Sources:	Part 1: 420 journals indexed; Part 2: 376 journals indexed.

Publishing Information

Provider:	*Content:* H.W. Wilson Company. *Electronic Presentation:* H.W Wilson Company.
Copyright:	H.W. Wilson Company.
Availability:	***This resource is available only to current faculty, staff and students of Columbia University.*** More information about LibraryWeb access

Searching

Search Methods:	Subject, author, title and other access points.

Related Resources

* After you have found citations, you will need to do a title search in CLIO to see whether Columbia owns this publication, for example, t=education update If you can't find it in CLIO, speak with a reference librarian.

Last Update: 30-jul-2001 13:52:28

Comments and Questions

APPENDIX B. Hierarchical Interface to LC Classification: Sciences

The Sciences section of HILCC (as of July 2000), constructed from portions of LC Schedule Q, S and G.

General Sciences	Q 0-390.999
Animal Sciences	SF 0-190.999, SF 191-275.999, SF 360.99-599.999, SF 600-1100.999, SH 0-400.999, SK 350-579.999
Astronomy & Astrophysics	QB 0-991.999
Biology (all)	*[composite of all Biology]*
General	QH 301-504.999, QH 506-539, QH 573-705.999
Biophysics	QH 505-505.999
Microbiology	QR 0- 502.999
Botany	QK 0-989.999
Chemistry (all)	*[composite of all Chemistry]*
General	QD 0-65.999
Analytical Chemistry	QD 66-145
Biochemistry	QD 415-449
Crystallography	QD 732-999.999
Inorganic Chemistry	QD 146-97.999
Organic Chemistry	QD 198-414
Photochemistry	QD 700-731.999
Physical & Theoretical Chemistry	QD 450-624
Radiation Chemistry	QD 625-699

APPENDIX B (continued)

Earth & Environmental Sciences (all)		*[composite of all Earth & Environmental Sciences]*
	Ecology	QH 540-572
	Environmental Sciences	GE 0-350.999
	Forestry	SD 0-669.999
	Geology	QE 0-996.599
	Human Ecology & Anthropogeography	GF 0-900.999
	Meteorology & Climatology	QC 851-999.999
	Natural History	QH 0-278.5
	Oceanography	GC 0-1581.999
	Physical Geography	GB 0-5030.999
Human Anatomy & Physiology (all)		*[composite of all Human Anatomy & Physiology]*
	Human Anatomy	QM 0-695.999
	Neurosciences	QP 351-495.999
	Physiology	QP 0-350.999
Mathematics (all)		*[composite of all Mathematics]*
	General	QA 0-74.999, QA 101-272.599, QA 299.5-939.999
	Mathematical Statistics	QA 273-299.499
Physics		QC 0-999.999
Plant Sciences (all)		*[composite of all Plant Sciences]*
	General	SB 0-450.899, SB 599-998.999
	Agriculture	S 0-972.999
Zoology		QL 0-991.999

APPENDIX C. Library of Congress Schedule Q (Selections)

A selection from LC Schedule Q showing detail corresponding to
HILCC Science categories.

General Science	Q 1-390
Mathematics	QA 1-939
Astronomy	QB 1-991
Physics	QC 1-999
Meteorology. Climatology	QC851-999
Chemistry	QD 1-999
General (including alchemy)	QD 1-65
Analytical Chemistry	QD 71-142
Inorganic Chemistry	QD 146-197
Organic Chemistry	QD 241-441
Biochemistry	QD 415-436
Physical and Theoretical Chemistry	QD 450-801
Radiation chemistry	QD 625-655
Photochemlstry	QD 701-731
Crystallography	QD 901-999
Geology	QE 1-996.5
Natural History. Biology	QH 1-705.5
General Natural History	QH 1-278.5
General Biology	QH 301-705
Biophysics	QH 505
Ecology	QH 540-549.5
Botany	QK 1-989
Zoology	QL 1-991
Human Anatomy	QM 1-695
Physiology	QP 1-981
General	QP 1-345
Neurophysiology and Neuropsychology	QP 351-495
Microbiology	QR 1-502

University of Washington Libraries Digital Registry

Kathleen Forsythe
Steve Shadle

SUMMARY. In 1998, the University of Washington Libraries transferred point of delivery for online services to the Web. Part of this transfer involved the design and implementation of a database of electronic resources, called the Digital Registry. Web resources are cataloged using existing workflows and data elements are transferred from the MARC record in the catalog to an SQL database. Records are mapped to subject categories by Library of Congress classification and organized within this by resource types. This article examines design considerations, workflows, maintenance, and usage of the resulting product. *[Article copies available for a fee from The Haworth Document Delivery Service: 1-800-HAWORTH. E-mail address: <getinfo@haworthpressinc.com> Website: <http://www.HaworthPress.com> © 2002 by The Haworth Press, Inc. All rights reserved.]*

Kathleen Forsythe is Electronic Resources Cataloging Librarian, Monographic Services Division, University of Washington Libraries, Box 352900, Seattle, WA 98195. She holds a BA in Anthropology and an MLS both from the University at Albany.

Steve Shadle is Serials Cataloger, Serials Services Division, University of Washington Libraries, Box 352900, Seattle, WA 98195. He holds a BA in Linguistics and an MLibr, both from the University of Washington.

The authors would like to thank the following University of Washington Libraries staff for use of screen shots from their Web pages: Figures 3A and 3B, Alan Grosenheider; Figure 4, Mark Carlson, and Figure 5 Patricia Carey.

[Haworth co-indexing entry note]: "University of Washington Libraries Digital Registry." Forsythe, Kathleen, and Steve Shadle. Co-published simultaneously in *Journal of Internet Cataloging* (The Haworth Information Press, an imprint of The Haworth Press, Inc.) Vol. 5, No. 4, 2002, pp. 51-65; and: *High-Level Subject Access Tools and Techniques in Internet Cataloging* (ed: Judith R. Ahronheim) The Haworth Information Press, an imprint of The Haworth Press, Inc., 2002, pp. 51-65. Single or multiple copies of this article are available for a fee from The Haworth Document Delivery Service [1-800-HAWORTH, 9:00 a.m. - 5:00 p.m. (EST). E-mail address: getinfo@haworthpressinc.com].

KEYWORDS. Web portals, organization of Web resources, cataloging electronic resources, Internet cataloging, electronic journals cataloging, University of Washington Libraries Digital Registry, Library of Congress classification mappings, metadata and subject access, MARC applications

BACKGROUND

In late 1997, the University of Washington (UW) Libraries began reengineering online service delivery from a locally developed X-Windows interface (Willow) to the Web environment. In addition to migrating locally mounted databases and the catalog to the Web, the project also included a redesign of the Libraries' Web site. In December 1997, a five-member prototyping team began work on reconceptualizing the site, which at the time was organized by the Libraries' administrative structure.

DESIGN CONSIDERATIONS/SYSTEM DEVELOPMENT

Catalog as Single Source

At the time of the redesign, three different access methods were provided for electronic resources. (1) There was the BRS/Search and Willow Z39.50 interface to abstracting and database services. Most of these services were locally-mounted, tape-loaded databases. (2) A few records for electronic resources were in the catalog although there was not yet URL access. The Libraries use Innovative Interfaces for technical services modules, but at this time used the same BRS/Search interface for the public catalog that was used to provide access to our other databases. (3) Subject specialists each maintained their own Web pages of links and there was no uniformity in how purchased resources of interest throughout the Libraries were presented on these pages. Some resources held in specific units were not known to be available for the whole campus. Some of these Web pages were also becoming larger than could be comfortably browsed or maintained.

The committee began by considering building an ejournals database which could support centralized maintenance and production of ejournal pages and could also facilitate links to library holdings and full-text. However, as deliberations proceeded, a central database was seen as a way to integrate all forms of selected electronic resources, whether paid or free. This database was named the Digital Registry and became a component of the Information Gateway. The Gateway serves as the main portal for the University Libraries and contains services and unit pages as well as the Registry.

An important design consideration was incorporation of existing staff and workflows. There was no budget for additional data entry. The catalog presented a logical central repository to take advantage of existing workflows both in cataloging and Library Systems, where there had been prior experience exporting records from Innovative to the local BRS catalog. There would be considerable overlap between resources made available through the catalog and the Digital Registry as well. So a single source of data input would avoid duplication of effort.

URL maintenance was another consideration. In the earlier environment, most URL maintenance was done by individual subject specialists on their subject pages with the result that resources appearing on multiple subject pages would require individual maintenance. An advantage of generating individual pages from a central database was that URL maintenance could be done once in a central database rather than on each individual page.

Elements Transferring from the Catalog to the Digital Registry

Because of the short development timeline, in-house tools needed to be used for implementation. MS-SQL, already available in the Libraries, was chosen as the database to drive the Digital Registry. The implementation group determined that a small amount of additional information needed to be added to the MARC catalog record to support access, processing and maintenance. Given our local cataloging environment, the group identified the check-in record as the place to hold most of this additional information and these fields were added to the Innovative check-in record structure. Fields include:

- E-access to designate whether the resource is unrestricted, IP restricted, requires a password, or is on the Libraries' CD-ROM network.
- URL was added to the check-in record even though it was already available in the bibliographic record. This duplication was necessary as there was no way to link a specific bibliographic record URL with the information in a specific check-in record when there was more than one online source for the same title.
- E-note allowing for the transfer and display of an annotation for end users of the Registry.
- Fields that do not transfer to the Registry, but are required for record maintenance. These include fields for contact email for staff to send requests for error correction, inputters' notes, and subject specialist/fund code of the resource.

In addition to these new elements in the check-in record, certain fields in the bibliographic record are transferred including date from the fixed field, ISBN and ISSN and other selected publisher numbers, language from 041 subfield a or the fixed field, selected call number fields, 043 geographic area code, 1XX, 24X, publisher from 260 subfield b, 520 and a few other note fields, 6XX and 7XX. Response time considerations precluded transferring the entire bibliographic record. Data elements for new and changed records are exported from the catalog and batch loaded to the SQL Digital Registry on a daily basis.

Organization

It was clear to the prototyping team from discussions, readings, and exploration of other sites, that no single navigation method would work well for all users. Therefore, a number of approaches to the same resources were built into the Information Gateway. The design for the top level Gateway page was task oriented, rather than organized by administrative structure. A need by individual units to produce journal lists of holdings suggested a subject approach. A search engine allows for searching of the Registry as well as the entire Information Gateway. Searching and browsing by author, title, subject and keyword is provided. Longer lists (e.g., "Databases and catalogs" and "Ejournals") contain an A-Z location jump bar. A feature called My Gateway allows end users to create their own custom interface to the resources they use the most.

To provide subject access, the Yahoo! approach of a broad category hierarchy was chosen as most appealing. By mapping Library of Congress classification numbers already assigned to bibliographic records to subjects, pages could be automatically generated and there would be no need to maintain static subject lists. Broad categories were determined based on disciplines in the curriculum and Libraries fund groups, and work began building the classification table to support these. It took several drafts of the hierarchy to satisfactorily map LC classification ranges to top level subject terms (see Figure 1).

In addition to the prototyping team, which consisted of systems, cataloging, and branch staff, several Libraries-wide committees were formed to work on different aspects of the migration to the new Information Gateway. The Web Content & Migration Subcommittee, among other tasks, created a list of electronic resource types. Examples of resource types include traditional bibliographic formats (e.g., bibliographies, catalogs, indexes) as well as Web-specific formats (e.g., electronic journals, Web sites, Web search engines). These resource types serve several purposes in the Information Gateway:

- they are the basis for the automatic generation of the Databases & Catalogs and Ejournals top level categories

- they serve as the organizational framework for automatically generated subject pages
- users may incorporate resource types into their searching.

The list was formulated after many discussions and has been altered very little in the ensuing three years (see Figure 2).

Each resource type was assigned an Innovative location code, so these provide an electronic resource "location" in the online catalog as well. Codes are added to check-in records and serve as the trigger for record transfer to the Digital Registry.

POPULATING THE DATABASE

The initial population of the Digital Registry was accomplished by a retrospective cataloging of links on existing subject pages. Resources were first divided by monograph vs. serial, as there are separate Monographic and Serial cataloging units. Within Monographs work is divided by subject area and lan-

FIGURE 1. Table of LC Class Mappings at http://www.lib.washington.edu/asp/registry/classranges.asp

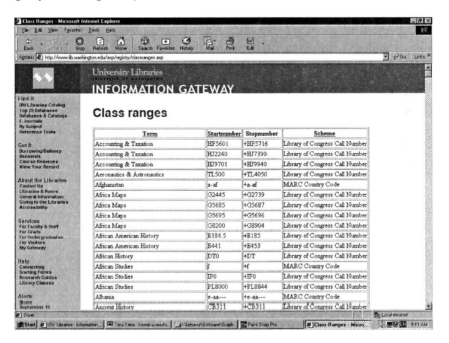

FIGURE 2. Chart of Codes for Electronic Resource Types Found at: http://www.lib. washington.edu/msd/type.html

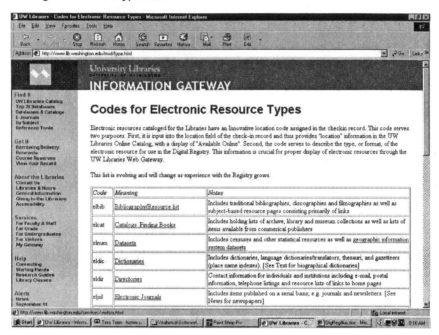

guage proficiency. Approximately a dozen catalogers were trained in creating records for Web sites, then proceeded with the conversion of pages within their areas of expertise.

The Serials Division (after consultation with public services staff) had already adopted the single-record approach for electronic serial cataloging. Since most of the electronic serials identified by subject specialists were already held in print, populating the database with electronic serials entailed editing existing bibliographic records and adding "digital registry" check-in records rather than creating catalog records. A short-term project staffed by one librarian and three serials staff processed nearly 2,000 electronic serials during Digital Registry development. One serials cataloger was assigned to catalog those serial titles not already held in print. Due to the tight deadline, this work was given top priority and by the time the Information Gateway went live in September 1998, approximately 5,000 monograph and serial titles were represented in the Digital Registry.

To accommodate ongoing submissions of new pages, a Web form was created (see *http://www.lib.washington.edu/asp/registry/add.asp*). The form contains pull down menus for type of resource and classes desired as well as name of subject specialist, fund code, URL, title and description. This assures that all the required information is collected to create a brief record in the catalog. The purpose of these brief records is to provide quick access to resources in the catalog and the Registry.

Messages from the Web form are directed to a dedicated e-mail address, which is checked at least once each working day. Requests are routed either to serials or monographic inputters who search the catalog for duplicates, OCLC for copy, and create abbreviated bib records and check-ins in the catalog. In Monographs, printouts of these records are sorted by subject and filed in folders on the distribution shelves to be picked up by catalogers. Catalogers then overlay the brief bibs with full cataloging. In Serials, acquisition staff handle the initial processing of these requests and forward those not already held in print to a serials cataloger.

RESULTS

Developing and populating the Digital Registry at the same time produced some unexpected results. For the first time, catalogers could assign more than one LC class number to a catalog record to bring out different aspects of the resource. The usual practice for monographs is to have the class number appear in the MARC 090 field. Class numbers for additional subject pages where this resource may also apply were added in 099s, the MARC field for local call numbers. While valid and interesting from a cataloging standpoint, in practice this led to problems. Individual subject specialists are responsible for particular pages, often using them for bibliographic instruction. Assignment of these extra call numbers resulted in resources that they had not selected displaying on their pages. The automatic generation feature causes a loss of control for what is displaying on a page as resources are added or deleted in the class mappings for that page. Policy for assignment of extra call numbers was adjusted so that it was only done on request from a subject specialist who wanted a particular resource chosen by another to appear on their subject page.

Once the general LC classification mapping was completed, subject specialists could define further mappings within their subject areas to refine their hierarchies. Some subject specialists chose to do this, others did not, which led to some variation in subject page display. Some chose to delineate further categories at the top of the page, sometimes creating an "Other" category to catch any resources not falling neatly into the top-of-page listings. The bottom of these pages often retain the automatically generated results for all pages in the LC class range mapped to the top subject in the hierarchy (see Figures 3A and 3B).

LC classification mapping does not work well with all units. International Studies units have a geographic rather than a subject focus, so 043 geographic

FIGURE 3A. Automatically generated page showing breakdown by subcategory. Any pages not mapped to a category display in Other Resources. See: http://www.lib.washington.edu/subject/International

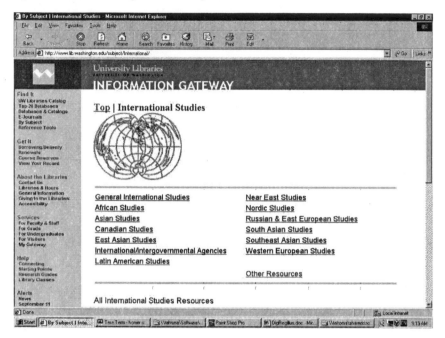

area codes were added into the mapping tables. Adjustments then needed to be made for those resources that were an unwelcome addition to a subject page based on the LC call number mapping while successfully displaying on International Studies pages from the 043. LC class mappings do not work well for units such as government publications and special collections and manuscripts, for interdisciplinary categories, or those subject specialists who simply prefer a different arrangement for their pages.

SOME SOLUTIONS

Resulting development efforts produced HTML editing solutions for those who did not want to use automatically generated pages. Subject specialists could substitute their own page for the default autogenerated page (see Figure 4).

They are then responsible for maintaining by hand the links on the page. Some chose to have a mixture of automatically maintained links and those

FIGURE 3B. Bottom of the page shown at Figure 3A. All resources are displayed in a single list organized by resource type. See: http://www.lib.washington.edu/subject/International

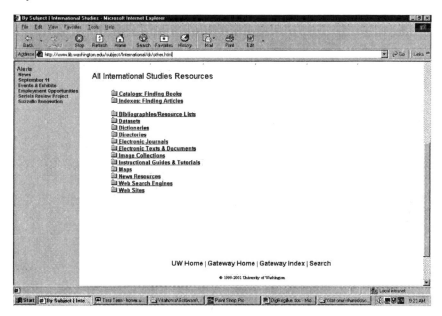

maintained by hand. In these cases, most often the ejournal link is automatically generated. This approach also allows for the incorporation of citations to print resources as well as electronic (see Figure 5).

The Databases & Catalogs category generated from the "elind" and "elcat" location codes contains all records coded as an index or catalog. It is significantly longer than the list of abstracting and databases services in the previous system. Public services reported some confusion on the part of users not sure of the best choice in such a long and diverse list. A separate handcrafted page was created of the Top 20 Databases to highlight the most comprehensive ones in various disciplines.

MAINTENANCE

At the time of the 1998 retrospective cataloging there were no selection guidelines for Internet resources in place. A committee completed work on selection guidelines for Internet resources in October 1999 (see *http://www.lib.washington.*

FIGURE 4. Entirely Handcrafted Page (http://www.lib.washington.edu/specialcoll/)

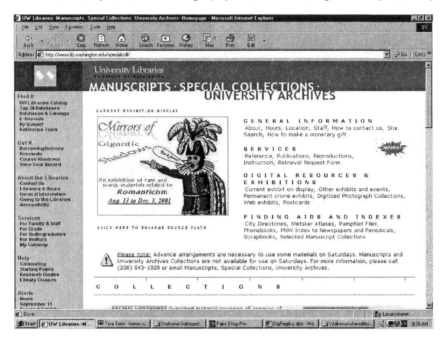

edu/msd/internetselguide.html). With these guidelines in place, a de-selection project began in December 1999. Lists were generated in the catalog by fund code and sent to subject specialists for review. Candidates for removal were posted for all to see in case others wanted to adopt cross disciplinary sites. In March 2000 over 600 records were removed from the catalog and the Registry.

For ongoing routine maintenance, a link checker runs in the Registry each working day. Broken links are written to a report when the URL fails for the third time. A procedure was developed and documented for handling broken URLs. Reports of broken URLs come in from staff and the public as well.

A steady stream of sites need to be recataloged because the title or content of the site have changed. These are usually discovered because the URL changes, titles merge, or the publisher changes. Whenever the content of a site changes significantly, the subject specialist is notified to review the site before it is recataloged. Title changes pose a particular challenge. There is currently no mechanism for cross references in browse lists in the Registry. Although most MARC title fields transfer and are searchable, only the 245 displays in browse lists. Often when a resource well known to users under a certain title

FIGURE 5. Handcrafted page using some automatically generated resource types. Electronic Journals is automatically generated. Selected Web Resources links to a handcrafted page that also contains automatically generated pages. Note the inclusion of a category on print resources (http://www.lib.washington. edu/subject/Environment/)

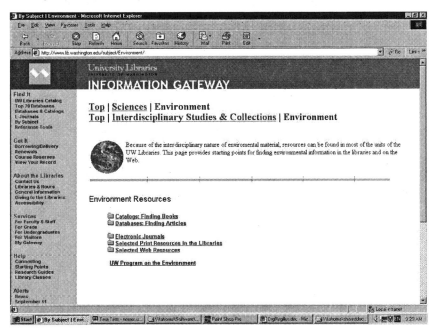

changes name, a brief bib is input into the catalog under the old title so there will be a cross reference.

Occasionally further work is needed on the classification mapping tables. Sometimes there is fine tuning of mappings or a subject specialist who wants to totally redo their subject pages. Currently there are 765 separate class ranges mapped to all terms in the Registry, including broad-level discipline terms (e.g., Humanities) and specific subject terms. A separate table relates the broader and narrower terms and is used to construct the first-level "By Subject" page listing the broad-level disciplines.

Individual catalog records sometimes need to be added or deleted from particular subject pages. 099 class numbers are added on request to records to "force" resources onto particular pages. Removing records from pages is more involved. As mentioned before, a single-record approach is followed as much as possible. If the hard copy of a resource is owned by UW Libraries, the URL is attached to this record rather than creating a separate record for the electronic resource. There have been a few cases where an unwanted resource appears on a subject page based on a clearly "correct" class number. Catalogers can delete these class numbers from the local bibliographic record, but given our BIBCO and CONSER workflows, there's the possibility that the class number will reappear in the local record if maintenance is being done on the record for some other reason. In these cases, we warn the subject specialist that the title might re-appear on a particular page and request that they ask us to delete it if it does reappear since our workflows don't allow for constant monitoring for call number deletion.

In following the single-record approach, historical cataloging practices and irregular cataloging data in print resource records can produce undesirable results in the Registry. For example, as part of a tape-load project decades ago with a different local system, the UW Libraries had records loaded with the print call number in the 099 field. When these records were included as part of the Digital Registry, some of these records had Dewey class numbers in the 099. The export program transferred the Dewey number to the Digital Registry record and mistakenly put these resources on a subject page based on the first letter of the cutter number. A validation routine could fix this problem. Since this was the result of a one-time project and call numbers indexed in the public catalog are stored in item or check-in records, it is not a problem to remove these from the bibliographic record on a case by case basis as a fix. As another example, pre-AACR2 serials cataloging regularly used the MARC 520 field to store issue-specific contents notes rather than a general summary, so data applicable to specific issues of the print version displayed in the Registry. A serials cataloger identified the records with this earlier MARC 520 practice and re-tagged the data as a 500 so that it wouldn't be exported to the Digital Registry.

In addition, there have been minor problems that have required further catalog record editing. For example, the decision was made that MARC 245 ab would be used as the display title for automatically generated listings. The unintended consequence of this was a list of 40 entries with the generic title "Annual report." The most expedient solution was to add the issuing body in a check-in E-note that would display next to the title. This editing was necessary since these titles were displaying not only as a group on the E-journal alphabetic listings, but were also appearing occasionally on subject pages with no other identification.

Until recently, there has not been formal instruction for newly hired subject specialists on the mechanics of creating and maintaining subject pages. As a result, there is unevenness in subject page development. There are also different philosophies of what should be in the Registry. Some favor presenting a few well chosen resources for their users. Others prefer in-depth lists. The

Web Steering Committee, which currently oversees development of the Information Gateway, recently started information sessions so that staff are aware of the tools and options available for subject page development.

LESSONS LEARNED

The lead time of approximately 5 months for planning and 4 months for implementation seemed too short for a project of this magnitude. Resources were cataloged without thorough review or criteria in place for selection. The database was created while programs were being written. There was little time for learning, implementation, or user testing on the part of public services before the Digital Registry went live.

The switch from individual subject specialist space to more of a shared space continues to be a challenge, with occasional adjustment of LC class numbers so that resources are added or deleted from pages. Some feel that a mapping of subject pages by fund code rather than by LC class number would have produced more satisfactory results. Some are not comfortable with pages automatically generated by class number and not knowing from day to day what resources may be showing on pages, so it is necessary to provide alternate ways of HTML page generation.

There are hazards as well as benefits to the reuse of data in more than one application. Materials cataloged according to AACR2 for the catalog, do not always yield the best title for browse lists in the Digital Registry. The classic example of this is the "Welcome to . . ." phrase, but there are other cases of resources known to users under a title that is not even on the Web page. Although time has been saved in centralized maintenance of URLs, there has also been much effort expended in tweaking data in various ways for better displays in the Registry.

Some workflow inefficiencies have been introduced as a result of working with local system functionality in ways outside of Innovative's original design intent. The main example of this is use of the check-in record as a source of data for the Digital Registry SQL database. This has not always yielded the most efficient workflows. For example, there was no time or funds to reconfigure the software loading OCLC records to the catalog to include electronic resource elements in the check-in record. During the initial database population, this meant that catalogers needed to go into the record in the catalog the day following OCLC input to add check-in elements that could not be loaded from the OCLC record. This happens less frequently now, since the check-in record is included in the local creation of the catalog records. A catalog record then overlays the brief bib record with the check-in record already attached. However, URL maintenance is still done in two locations, the bibliographic record and the check-in record, and often in the OCLC master record as well.

USAGE

A server snapshot taken on November 29, 2001 showed 279 subject "nodes," which include subject hierarchy pages at all levels. Eighty-six of these (approximately 30%) have been overridden by hand crafted pages. On this day, 2,226 HTML files (out of 3,342) in the /subject directory (where all of the subject pages reside) were modified indicating that about two-thirds of the total HTML files in the subject hierarchy pages are auto-generated.

Looking at "hit" statistics to the Library's Web server pages (lib.washingon. edu/) for 3rd quarter 2001, there were two automatically-generated pages which ranked in the top twenty: the Databases and Catalogs page ranked number eight with around 1.5% of the total visits to the Information Gateway. Ejournals, another automatically generated page from resource type ranked fifteenth with around 1% of the total visits. None of the individual automatically generated subject pages appear in the top 20 list. However, seven of the top twenty hit pages are the home pages for individual branch libraries. Because these pages are up and refreshed continuously on the public workstations in their respective library units, it is difficult to be certain that all of the hits represented here are actually associated with a real "use." Real usage of automatically-generated pages may be more than what is represented in the top documents category.

Because all of the subject pages are within the /subject directory, one can also examine hits within those directories compared to other directories on the Libraries' server to determine usage of automatically-generated subject pages. The directories report for this same quarter ranks the /subject directory third with nearly 7% of all visits. This is only behind the /proxy.pac and /ie5.pac directories which contain proxy software and support. Although the ranking of subject page usage in aggregate sounds good, when taking into consideration the number of pages represented in /subject, the per page usage is low in comparison to other Gateway pages.

Reinforcing this perception of low subject page usage is a Gateway survey conducted in the Fall of 2001 where only 2 of the 100+ comments addressed subject pages at all. Anecdotal evidence from public services suggests some user difficulties in navigating subject hierarchies. Various approaches have been discussed to help with this. One possibility is changing the "By Subject" link to go to the A-Z subject list rather than the hierarchies. An additional starting point for beginning researchers listing the 50 most comprehensive resources by subject has also been suggested, both to cut down on the number of clicks it takes to get to a resource and to present a shorter list of the best databases in each subject area. There is some concern that certain resources are getting overused because of their prominence in the Top 20 list when there are other discipline specific resources more pertinent to user needs underutilized because they are buried in subject hierarchies or long lists. Additional user testing is planned to gather more feedback.

By the end of November 2001, there were 18,940 records in the Digital Registry. Most of these records are in the catalog as well, with a few exceptions such as UW department pages being just in the Registry. There is ongoing discussion and differing opinions as to which resources should be in the catalog and which in the Registry. A recent trend is the proliferation of e-resource records from purchased sets that analyze titles on megasites such as NetLibrary, or purchased or locally produced e-journal aggregator sets such as Proquest and Lexis/Nexis. Due to workflow and maintenance issues, only a small number of these sets have been loaded into the Registry, so increasingly the catalog is becoming the most complete source for e-resource discovery.

FURTHER READING

William Jordan, "My Gateway at the University of Washington Libraries," *Information Technology and Libraries* 19, no. 4 (December 2000): 180-185.

Digital Registry presentations are available at: *http://www.lib.washington.edu/about/registry/*.

Bridging the Gap Between Materials-Focus and Audience-Focus: Providing Subject Categorization for Users of Electronic Resources

Jonathan Rothman

SUMMARY. Traditional subject cataloging is based on the content of individual items without reference to user context. Hierarchical, browsable Web-based lists, using institutionally meaningful subject taxonomy, would provide desirable access for many users. Lists could be produced without double-maintenance of data if the local subject terms can be mapped from LC classification numbers already in catalog records. Creating such a mapping system poses many challenges, but also holds great promise. *[Article copies available for a fee from The Haworth Document Delivery Service: 1-800-HAWORTH. E-mail address: <getinfo@haworthpressinc.com> Website: <http://www.HaworthPress.com> © 2002 by The Haworth Press, Inc. All rights reserved.]*

KEYWORDS. Subject categorization, mapping, audience-focus, materials-focus

Jonathan Rothman is Senior Systems Librarian/Analyst, University of Michigan University Library, Ann Arbor, MI 48109 (E-mail: jrothman@umich.edu).

The author would like to acknowledge Paul Schaffner, of the University of Michigan University Library, whose analysis forms much of the basis for this article.

[Haworth co-indexing entry note]: "Bridging the Gap Between Materials-Focus and Audience-Focus: Providing Subject Categorization for Users of Electronic Resources." Rothman, Jonathan. Co-published simultaneously in *Journal of Internet Cataloging* (The Haworth Information Press, an imprint of The Haworth Press, Inc.) Vol. 5, No. 4, 2002, pp. 67-80; and: *High-Level Subject Access Tools and Techniques in Internet Cataloging* (ed: Judith R. Ahronheim) The Haworth Information Press, an imprint of The Haworth Press, Inc., 2002, pp. 67 80. Single or multiple copies of this article are available for a fee from The Haworth Document Delivery Service [1-800-HAWORTH, 9:00 a.m. - 5:00 p.m. (EST). E-mail address: getinfo@haworthpressinc.com].

Subject cataloging and classification have traditionally provided access to materials based on content, in what can be called a "materials-based" approach. Subject catalogers assign subject headings and classification numbers, from a very large potential set, based on analysis of the content of each specific work. Headings and numbers are assigned based on the individual item considered in isolation. The vocabulary is not adjusted based on the context from which a particular user or set of users operates. The library literature is replete with studies indicating that many users do not understand subject headings well. One of the issues frequently identified is a mismatch in vocabulary. Collantes, in a study of how students (undergraduates and graduates) and librarians approached subject classification, concluded that ". . . there is little agreement in the names that people use and the names recommended by LCSH" (Collantes, 1995, 130).

Outside the library realm a recent study of how users categorize Web sites identified two groups of decision factors: "content attributes," i.e., those attributes that are intrinsic to the material being categorized, and "context attributes," which affect classification ". . . in terms of the individual's relationship to the information itself. . ." (Gottlieb and Dilevko, 2001, 524). While "topic" is classified as a content attribute, the study suggests that it also functions as a context attribute because users interpret the meanings of topics differently. Once again, there is a vocabulary mismatch problem–users do not share a common vocabulary for the terms chosen to describe topics.

While catalogs have traditionally provided the primary subject access to library collections, many libraries have also provided another type of topical access to selected materials through the creation of bibliographies and pathfinders. These lists provide access to materials classified according to one or more broad topics that are deemed to reflect areas of interest for specific groups of users. Works are assigned to topics that are named and defined based on users' perceived interests and perspectives. The topics used and the lists produced are based, at least in part, on the list creator's knowledge of the content in relation to the information vocabulary and needs of a group of users. Thus they are informed by context attributes and can be seen as "audience-based." In general, the traditional paper audience-based tools that libraries have created have not been intended to provide comprehensive access to collections, and they have been produced independently of the library catalog by entirely different staff.

There appears to be ample evidence that many of today's users are most comfortable accessing materials through hierarchical Web-based lists of subjects like Yahoo! Antelman notes that "Internet searchers value collocation by subject, of course, and Yahoo!'s readily browsable classification system has made it the most popular site on the Web" (Antelman, 2000, 190). Many libraries' initial responses have been to produce online bibliographies and pathfinders using the same methods they have used to produce paper tools: individual subject specialists and bibliographers manually create and maintain lists that reflect small selective slices of the collection. This approach is not viable in the current environment for two reasons:

- First, individual selectors, working manually, cannot keep up with the burgeoning amount of material available on the Internet. In a survey on the creation and maintenance of Web-based subject guides, many librarian respondents reported their schedules for updating their pages as: "Whenever librarians get around to it" or "During breaks as I get to it" (Morris and Grimes, 2000, 214).
- Second, users want and expect hierarchical audience-based access to more than selected slices of our collections. They expect browsable, hierarchical interfaces that organize access to full collections using terms that they find meaningful.

To meet these user demands we need a way to produce Web-based lists that are up-to-date, comprehensive, and maintainable. This cannot reasonably be accomplished through manual list creation and maintenance. We cannot afford to create, acquire, and maintain the same data in multiple, independent streams. One attractive approach is to generate these lists using the electronic metadata we already use to build our online catalogs, but this raises a large issue. We want our Web-based lists to reflect an audience-based subject vocabulary, but the subject access available in our source data, the catalog, is materials-based. How do we bridge the gap from our materials-based data to the audience-based access we wish to provide?

In the following pages, I will describe the approach the University of Michigan University Library has taken toward providing linkage between content and audience-based subject access tools to date, and our thoughts and preliminary research on next steps and new directions.

RECENT HISTORY

As with many of our peer institutions, the first audience-based electronic lists of resources and pathfinders at the University of Michigan were produced according to the traditional paper model. They were written and maintained by individual subject specialists to meet the perceived needs of a particular user population. As such, the contents were either classified within a single topic covered by the list or within a few hand-selected topics. Lists had no connection to the online catalog or the full cataloging records it contains. Not surprisingly, the authors of these pages found maintenance to be burdensome and the pages tended to go quickly out-of-date and become out-of-synch with the data available in the online catalog. This became especially evident when user demand for a browsable Web-based list of electronic journals began to grow.

In June of 1998, the Access to Electronic Resources Task Force (AccessER) was formed. Among the group's initial charges was to create a comprehensive, maintainable Web-based list of electronic journals available to University of Michigan users. By that point there had already been a good deal of thought and discussion in the library literature on the pros and cons for including electronic resources in library catalogs. The AccessER group concluded that, in our case, including internet-accessible and other electronic titles in our catalog represented our best available tool for managing and providing access to the electronic resources that had grown to be a significant part of our collections.

We also concluded that the best option for creating a unified electronic journals list in our environment was to produce the list entirely from data residing in the catalog records. This eliminated the need for parallel maintenance and enabled the electronic journals list to be kept up-to-date and in-synch by fully replacing its content from the catalog on a regular basis. These decisions had two significant implications:

- All electronic journals that we collected needed to be represented in our catalog if they were to be included in the list. As a result of this decision, our serials cataloging department stepped up cataloging for electronic journals and began a massive process of analyzing the titles included in large vendor and aggregator packages to which we subscribe.
- We needed a subject or topical access scheme that would be meaningful to our users in a browsable list (in other words, we needed an audience-based subject scheme using a vocabulary that would be meaningful to our users) and we needed a way to apply that scheme to the catalog records which were to be the source for our electronic journals list.

Beginning with a list of topics drawn from several manual lists produced by other universities, we made use of course catalogs and wide-ranging input to create a list which approximated the department structure of the University of Michigan. Not surprisingly, creation of this list was a politically charged process, with various constituencies advocating for additional more specific representation for their subject areas. Countering that push was our recognition that an overly large set of topics would greatly reduce the usability of the electronic journals list. Eventually we settled on a set of about 55 topical headings plus sub-headings for geographic areas and the publication languages of newspapers.

We then were faced with the task of associating these topics with the records for electronic journals in our catalog. Given our goal that all data maintenance should occur in one place (the catalog), our Serials Cataloging Department took on the daunting task of adding one or more of our newly defined topics as local subject headings (in MARC 690 fields with a locally-meaningful first indicator) to all existing and future electronic journal records. While this approach has resulted in a product that is very well received and heavily used, it has become more and more difficult to sustain in the ensuing period as the number of journal titles represented has grown from about 1,500 to over 13,000. Beyond the huge time commitment involved in initially adding local subject headings to the records, we have the inevitable problem that growth in the list is not evenly distributed amongst all topics. Some topics have grown to include many hundreds of titles, far too big for a browsable list. While our initial approach served our needs at the time, it is very inflexible, since splits and changes to the topical scheme require editing the local subject headings in large numbers of individual records.

BRIDGING THE GAP

In recent months we have begun to confront two major issues associated with the manner in which we provide audience-based subject access to electronic resources:

Proliferating Subject Schemes

Separate subject taxonomies have been constructed for products and access tools (e.g., electronic journals list, networked electronic resources list, new books list, etc.) that were developed in different ways and at different times. There is fairly broad consensus that this is confusing for users and staff and that a single common vocabulary that can be used across products would provide a better service. Defining a single vocabulary is, however, greatly complicated by legitimate differences between the access tools and the materials they classify. For example, our Networked Electronic Resources List remains a relatively small, selective listing, which provides access to full-text and reference databases that are created, converted, owned, or licensed by the University of Michigan. It includes many large, multi-disciplinary databases, such as Lexis Nexis Academic Universe, with very broad subject coverage. Our new books list and electronic journals list, however, contain considerably larger numbers of titles, many with highly specific subject coverage.

To address these issues, we are proposing a two-level subject hierarchy with the top level providing a set of topics suitable for the Networked Electronic Resources List, and a second level that is loosely based upon University Schools and Departments, developed from the existing set with input from a review team. The second tier numbers about 100 topics, distributed under one or more of the more general top level categories. We have deliberately tried to limit the hierarchy to two levels, feeling that a third is too many layers to ask users to navigate and that the complexity of identifying where a chosen topic lies would be difficult to display in a comprehensible manner.

Once again, settling on the specifics of the topic list is likely to be a highly political process involving significant negotiation. Details of that negotiation, and the lists themselves, are outside the scope of this discussion but it is important to note that the two-tiered subject model will add complexity to plans for automatic subject term generation.

Automatic Subject Term Generation

If the products we are developing are to be maintainable, it is clear that we need to develop a method for automatic generation of our audience-based subject terms based on the data residing in our catalog records. Doing so will allow us to eliminate manual entry and maintenance of the local subject headings currently used for this purpose. In other words, we need to map the materials-based data already in our catalog records to the set of audience-based topics we ultimately agree on. When such a mapping scheme has been implemented, changes in the set of audience-based topics can be addressed largely through modification of the map(s) rather than modification of local subject headings that reside in each individual catalog record.

During the past year, AccessER began to discuss and study options for automated mapping from materials-based to audience-based subject classification. At the outset we made two assumptions: that a significant portion of our collections contain LC call numbers which can be used as source data for mapping; and that developing maps from LC call numbers to audience-based topics will take significant work on the part of both subject specialists, who understand the scope and meaning of those topics in our environments, and from catalogers, who understand the scope and content of the LC classification schedules. We also recognized that many of our bibliographic records for electronic works do not currently contain call numbers (since they cannot be shelved), and that those that do may not always contain the set of call numbers needed to successfully implement mapping. Finally, we acknowledged that mapping from call numbers to broad topics is inherently imperfect and will inevitably exclude some relevant materials and include some extraneous materials.

TESTING

In light of these acknowledged assumptions and limitations, AccessER began a pilot to further explore issues and questions about mapping from LC call numbers to a local audience-based subject scheme. This pilot had two goals: (1) to collect some preliminary data on the potential success rates that might be achieved through such a mapping process; and (2) to experiment with developing the needed maps and, in doing so, provide some basis for estimating the kind and amount of effort that might be required to develop our "topic maps."

Two subject areas were chosen to test the concept. In the first pilot, we looked at the area of African and African-American studies. In this case the subject specialist for that area worked together with an experienced cataloger to develop a draft topic map. Pooling their knowledge to identify starting points, they began a series of iterative keyword searches in the LC classification schedules using the LC *Classification Plus* searching interface. In the course of two meetings, totaling approximately four hours, they identified a list of 242 relevant call numbers, call number stems, and/or call number ranges. Both participants report that working as a team was an efficient way to generate the list. Once this draft topic map was compiled, we began to test how effective it would be in retrieving appropriate electronic journal titles.

Because the actual mapping mechanism has not yet been developed, we needed to simulate the map results through other means. We approached this task in the following way. Our systems office programmed and ran a special report that identified all of the serial titles (from the electronic materials subset of our catalog) that contained one of the classification numbers from the draft topic map. Records were considered to match both where a class number occurred as an active call number in use in our system and where it occurred in a bibliographic record call number field (e.g., MARC 050 or 090) but was not used as a call number in our system. This initial selection process identified 57 titles that matched a call number in the topic map.

Next a manual process was undertaken to extract a list of relevant titles from two subject areas in our current electronic journals list: "Cultural/Ethnic Studies" and "Areas of the World/Africa." (This manual extraction was required because the African/African-American studies subject is a part of our new proposed two-tier hierarchy, but it does not exist in the current manually entered scheme.) The total number of titles listed under those two existing categories numbered 321. Of that number, 103 titles were clearly relevant to African and/or African-American Studies.

Examining the discrepancies, we found that there were 44 titles that appeared on both the hand-selected and machine-selected lists. This left thirteen titles selected by the automated process that did not appear on the manual list and 59 titles that were selected by hand but not retrieved based on our initial topic map (see Table 1). Additional manual analysis was then undertaken in order to identify the factors leading to these discrepancies.

The thirteen titles that were selected by the automatic process but were missed by the manual process could all be explained relatively simply. None of them was included in the current electronic journals list. Of these, about half probably should have been included based on their topical coverage, but the local subject heading which would have caused their inclusion had not been added to the catalog records. The other half were a group of general-interest children's magazines that the subject specialist chose for the topic map, but which were not specifically African-American in coverage and, thus, had not been identified for inclusion in the existing electronic journals list. From this we concluded that the initial map did not appear to overselect except where it was specifically requested to do so (i.e., the children's magazines).

The 59 titles selected by the manual process, but not by the initial map, were found to have a variety of explanations (see Figure 1):

- Thirty-five of them had no call numbers at all, but could have. Of this number, twenty-nine were newspapers, which have not traditionally received call numbers in our system.
- Eleven of the records had call numbers that were not included in the draft topic map but which were appropriate to the topic and probably should be added to it.
- Seven of the titles had call numbers that emphasized another aspect of the item's subject matter (e.g., "US Black Engineer" is classified under civil engineering).
- Four of the titles referenced Africa only in a cutter number.
- Two of the titles were of more general interest and had general interest classification numbers in their bibliographic records.

This breakdown leads us to several conclusions on the results of potential improvements to the topic mapping process (see Figure 2):

1. There were 110 titles in the catalog that could have been legitimately selected based on the topic "African and African-American Studies." A revised topic map that added a few additional classification numbers would have selected 62 of them (56%).
2. Another 35 (32%) of the appropriate records could be retrieved by adding call numbers to titles that currently lack them, especially to records for newspapers. Addressing this issue would involve both retrospective work to add call numbers to records which currently don't contain them as well as policy changes and prospective work to add call numbers to

records for materials where practice has not been to assign them (e.g., newspapers and records for materials available only via electronic means where call numbers have been considered unnecessary since, by their nature, they do not have shelf locations).

3. Another eleven (10%) of the possible titles would be retrieved by adding legitimate alternative class numbers to their bibliographic records, bringing retrieval to approximately 98%. Doing this would also likely result in significant additional cataloging work, both retrospective and prospective. Success would require either that catalogers were able to accurately predict which class numbers would need to be added to match the existing topic maps, that there be an iterative process which identified missed titles so that additional class numbers could be added or, likely, both.

4. The remaining two titles (approximately 2%) seem unlikely to be retrieved based on a class number map at all.

This analysis work was conducted by an experienced cataloger who had not been involved in definition of the draft map. Completion of the analysis took roughly 10-12 hours of work. While efficiencies can no doubt be achieved through better-defined map creation, prototyping, and refinement processes, much time-consuming manual map analysis and refinement work is likely to be required for each topic map that is developed.

In the second pilot case the Philosophy subject specialist, after an initial strategy consultation with a cataloger, has worked by himself to develop the draft topic map. While considerable time and effort has been expended on this work, it is not yet complete. Based on progress to date, he currently estimates that total time to complete the map will run 25-30 hours. One factor identified as a major contributor to the length of this process is the diffuse nature of the topic. This has led him to perform keyword searches for fairly generic terms (e.g., "philosophy" and "theory") throughout all of the classification schedules. These searches produce large numbers of results that need to be evaluated for relevance. In many cases, he has chosen to examine numerous volumes from the shelves in order to clarify the relevance of a particular classification number. His goal has been to produce a comprehensive topic map in the first draft.

TABLE 1. Comparison of Selection Results Between Initial Topic-Map and Manual Process

Process	Total Titles Selected	Titles Selected by both methods	Unique Titles Selected
Topic Map	57	44	13
Hand Selection	103	44	59

FIGURE 1. Titles Selected Manually but Not by Initial Map

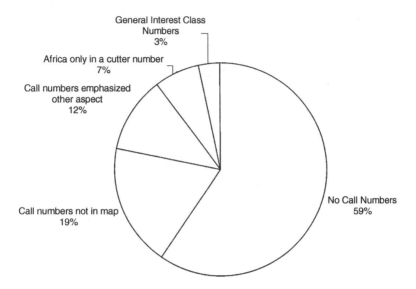

While this work is not complete, we can draw several tentative conclusions from this pilot that can help to shape our overall methodology:

- It is likely more efficient for a subject specialist to develop topic maps in concert with a cataloger, who is deeply familiar with classification schedules, than it is to develop the maps alone.
- It is probably quicker to produce an initial draft map that may have gaps, which can be corrected through iterative testing, than it is to attempt full comprehensiveness on the first draft.
- There will be significant variations in the difficulty (and thus time required) to produce maps for different topics.
- Even where significant commitment to the goals of the project exists, identifying and prioritizing adequate time to complete this work by a very large number of staff will be a major challenge.

PILOT ASSESSMENT

While it is clear that there will be significant variations amongst specific topics, if we assume that results for other topics would show some similar patterns

FIGURE 2. Results of Potential Improvements to Initial Topic Map

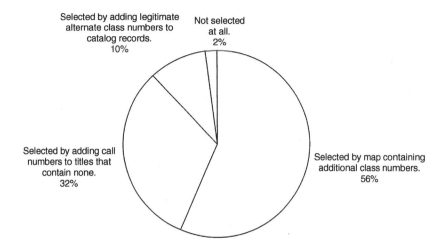

Selected by adding legitimate alternate class numbers to catalog records. 10%

Not selected at all. 2%

Selected by adding call numbers to titles that contain none. 32%

Selected by map containing additional class numbers. 56%

to the African/African-American Studies case, we can draw some general conclusions about implementation of the mapping process on a broader scale:

First, development of successful topic maps will require multiple, iterative steps many of which will involve significant effort:

- Joint work between subject specialist and cataloger will be required to develop a solid draft topic map in an efficient manner. In the African/African-American Studies pilot, this required about four hours work. Some topics may take less time. Many will take longer. While some efficiencies can no doubt be introduced, developing maps for approximately 100 topics represents a significant time commitment.
- Prototype implementation of the map and a methodology for verifying the accuracy of retrieval are needed. It may be possible to automate parts of this process, but inevitably it will require significant human intellectual effort to assess whether the correct materials are being retrieved. In the pilot, this work required approximately 10 hours of work. Even if that could be substantially reduced, performing this work for approximately 100 topics will require a large amount of staff time.
- There will need to be one or more rounds of tuning for each map by adding and/or changing the class numbers included to address identified deficiencies. We do not yet have time and effort estimates for this work.

Second, a significant portion of the relevant titles will only be retrieved by retrospectively adding class numbers to the bibliographic records for titles that are identified as missing from the retrieval set. This retrospective modification of records is another significant time commitment.

Third, successful retrieval will also require changes to policy and prospective cataloging practice to add class numbers to records which would not have otherwise been classified and, possibly, to add additional class numbers to records that would have been classified more narrowly. Policy discussions typically involve significant time commitments from several people. Changes to prospective workflow, especially if it leads to changes in copy-cataloging records that might not have otherwise been touched are also likely to have a significant impact.

A common thread running through all of the above is the significant time commitment required to implement topic mapping. Discussions of the costs and benefits that can be achieved are ongoing at the University of Michigan.

CLASS NUMBER MAPPING IS NOT EXHAUSTIVE

Even if all of the above is implemented, our pilot (and common sense) tells us that retrieval via topic maps will not be exhaustive. Some of the material which might be considered relevant will not be retrieved by the map. Three of the many possible approaches to this issue are laid out below:

- Acknowledge on an institutional level that the results are imperfect but agree that they are good enough to live with. We need to remember that manually produced lists and bibliographies are, inherently, not perfect or complete and that neither catalog searching, nor any of our other previous tools, has ever produced exhaustive results. Even if we agree to live with imperfection, however, it seems clear that we would need to create a method for forcing specific titles, reported by patrons and subject selectors as missing, to map into particular categories. This might be accomplished most simply by adding additional class numbers to the catalog records for those titles. One possible issue that may arise is that the class numbers that would most effectively improve mapping would not always be defensible from a cataloging point of view. A policy addressing such issues will be needed.
- Develop a process that would allow specific records to be marked for inclusion in topic maps, independent of the class numbers to be included. Leaving aside any interface development work that would be required, this implies either: that subject specialists would be directly updating the

catalog records for these titles with local mapping information in some as-yet-undetermined form and location; or that title-specific local mapping information would be stored in a database independent of the catalog data itself. While, in certain respects, this last may be the most appealing choice, it violates one of our basic principles by re-introducing multiple data-sources that must be kept synchronized.

- Supplement the mapping from classification numbers with limited mapping from Library of Congress Subject Headings. Based on our African/African-American studies pilot case, mapping from fourteen subject headings and/or subdivisions would have retrieved all but one of the relevant titles that were not selected based on class mapping. While we have not completed controlled testing based on the addition of subject headings to the topic map, it seems inevitable that we would be improving recall at the cost of precision. Some of the needed subject headings and subdivisions are quite general in nature (e.g., 650 subfield |z Africa) and may well result in adding titles to the list that might not be desired.

IMPLEMENTATION ISSUES

The previous discussion has focused on the political and intellectual issues involved in defining the content of topic maps. There are also, of course, a number of concerns related to the technical implementation of topic mapping. Given our intent to provide a common user-based vocabulary across various interfaces and products, map implementation will need to be independent of any individual system. Preliminary plans are relatively simple. The maps would be loaded into a relational database that can be queried from within each of the processes used to build and refresh the access products and tools.

Effective design of the database tables will depend on designing another type of common vocabulary. We need a clear, unambiguous syntax in which to express the classification numbers used to build the maps. For instance, does a single number function as a stem, matching on that string with any number of trailing characters, or does it require an exact match? Defining this vocabulary in a clear and flexible way will depend on joint work by catalogers, who understand the ramifications of matching on classification numbers in different ways, and systems staff, who understand how different matching algorithms are likely to affect complexities in the system design and efficiencies in its use.

In addition, given the identified need for iterative editing when establishing new maps and the likelihood of changes to maps over time, it will be highly desirable to provide an easy-to-use interface that will allow subject selectors and/or catalogers to modify maps directly without the need to involve systems staff.

While not absolutely necessary at the beginning of a large-scale topic-mapping project, such an interface is likely to significantly improve staff acceptance of the mapping system.

CONCLUSION

Based on the preliminary work and planning carried out at the University of Michigan to date, it appears that mapping from LC classification numbers could potentially select and categorize a high percentage of relevant materials into a user-based subject taxonomy. This will allow automated generation of access tools in forms that many users desire, while source data can be maintained in a single place (the catalog). It is also clear that developing such a mapping scheme on a broad scale will require a significant commitment of staff resources from across the institution to be successful. In sum, there will be significant costs to bridge the gap between materials-based and audience-based subject access, but there is a high potential payoff for those who choose to embark on this work.

REFERENCES

Antelman, Kristin. 2000. Web lists and the decline of the library catalog. *Library Computing* 18 (3):189-195.

Collantes, Lourdes Y. 1995. Degree of agreement in naming objects and concepts for information retrieval. *Journal of the American Society for Information Science* 46 (2):116-132.

Drabenstott, Karen M., Schelle Simcox, and Marie Williams. 1999. Do librarians understand the subject headings in library catalogs? *Reference and User Services Quarterly* 38 (4):369-387.

Gottlieb, Lisa and Juris Dilevko. 2001. User preferences in the classification of electronic bookmarks: Implications for a shared system. *Journal of the American Society for Information Science and Technology* 52 (7):517-535.

Marmion, Dan. 2001. Library web page design: Are we doing it right. *Information Technology and Libraries* 20 (1):2-3.

Morris, Sara E. and Marybeth Grimes. 2000. A great deal of time and effort: An overview of creating and maintaining internet-based subject guides. *Library Computing* 18 (3): 213-216.

Competing Vocabularies
and "Research Stuff"

Keith A. Morgan
Tripp Reade

SUMMARY. The NCSU Libraries portal, MyLibrary@NCState, presents customization and personalization options to students, faculty and staff. These profiles must be constructed in such a manner that eliminates unnecessary complexity and library jargon. At the same time, this process must also represent the unavoidable complexity of the research process. This paper examines the discussion that underlies the profiling and resource description framework and highlights some procedural and political problems that designers of library portals must take into account. *[Article copies available for a fee from The Haworth Document Delivery Service: 1-800-HAWORTH. E-mail address: <getinfo@haworthpressinc.com> Website: <http://www.HaworthPress. com> © 2002 by The Haworth Press, Inc. All rights reserved.]*

KEYWORDS. Portals, customization, personalization, social communication, vocabularies, jargon

Keith A. Morgan is Client Services Librarian, Digital Library Initiatives Department NCSU Libraries, North Carolina State University, Raleigh, NC (E-mail: keith_morgan@ ncsu.edu).

Tripp Reade is Media Resources Librarian, Access and Delivery Services Department, NCSU Libraries, North Carolina State University, Raleigh, NC (E-mail: tripp_reade@NSCU.edu).

The authors dedicate this article to the student who told them, "I'm a history major, but I don't like doing all that research stuff."

[Haworth co-indexing entry note]: "Competing Vocabularies and 'Research Stuff.' " Morgan, Keith A., and Tripp Reade. Co-published simultaneously in *Journal of Internet Cataloging* (The Haworth Information Press, an imprint of The Haworth Press, Inc.) Vol. 5, No. 4, 2002, pp. 81-95; and: *High-Level Subject Access Tools and Techniques in Internet Cataloging* (ed: Judith R. Ahronheim) The Haworth Information Press, an imprint of The Haworth Press, Inc., 2002, pp. 81-95. Single or multiple copies of this article are available for a fee from The Haworth Document Delivery Service [1-800-HAWORTH, 9:00 a.m. - 5:00 p.m. (EST). E-mail address: getinfo@ haworthpressinc.com].

On the treeless tundra of the high Arctic, the native Inuit have for centuries constructed piles of stone into formations of varying size and shape. These stones are Inuksuit, the "silent messengers of the Arctic." In Inuktitut, the Inuit language, Inuksuit are said to "act in the capacity of a human."[1] These stones serve as directional markers or signposts. Some indicate the presence of game or point the way home or celebrate life. The most familiar structure, an Innunguaq, mimics the shape of a person with arms or legs outstretched. Another common type of Inuksuit, the nalunaikkutaq, or "deconfuser," reminds the builder of previously cached equipment or an appointment with another person.[2]

The Inuksuit convey information to an Inuit–for example, distance, location, or availability of food or water. If to an uninitiated stranger an Inukshuk–the singular of Inuksuit–is merely an aesthetically pleasing pile of stone; to the Inuit, an Inukshuk is a sign that conveys vital information. As knowledge workers understand, information, even in the guise of a pile of rocks, is often incredibly complex. The philosopher Andrew Borgmann thinks that at least five factors must be present, suitably related and instantiated, in order to produce information:

> INTELLIGENCE provided, a PERSON is INFORMED by a SIGN about some THING within a certain CONTEXT.[3]

For Borgmann the sign is "the fulcrum of the economy of information." But he accurately observes that "the technology of information has loosed a profusion of signs" and as a result "information is about to overflow and suffocate reality."[4]

To suffocate reality, information overflows and produces "information overload." What happens in this phenomenon is that one's capacity for sustained concentration, attentional reserves, and ability to choose become exhausted. Two of psychologist Warren Thorngate's principles of attentional economics help describe what happens to a user during the course of a search: that attention is a finite resource, it can be focused on one activity at a time, and that the very act of searching is itself a drain on this resource.[5] Picture the average desktop, with its myriad open windows, add in the hyperlinked hopscotch that characterizes any search through a digital environment such as the Internet, and the result is a centrifuge perfect for whirling attention to the four winds. As David Levy comments, "To spend our limited attentional budget well, we must make wise choices, but the act of choosing will cost us too, sometimes dearly."[6] When designing, refining, and maintaining a portal, it behooves us to remember that the research process itself will drain the user; they don't need a portal that makes the process more difficult.

Of the many locations where this overflow and suffocation of information are occurring, none is more poignant than the case of the university library. Certainly all libraries are affected by this exploding universe of digital information access and retrieval. Nevertheless, the university library is, for a majority of students, the ultimate, and often final, environment in which to confront the world's collected intellectual, cultural, and scientific wisdom. The difference is that this confrontation is occurring in a significantly changed environment from what it was a mere ten years ago. To tame this new digital environment new tools are needed.

Lest there be any doubt that users are clamoring for change, consider the recent LibQUAL+ survey. LibQUAL+ is an ARL initiative to define and measure library service quality; users rank their library in terms of minimum and desired levels of service on a number of items. This gives a "zone of tolerance." Users then register a third, crucial ranking: perceived level of service. An institution using this survey hopes the perceived level of service lies close to the desired level. Of the four constructs measured, "Personal Control," which measures a university's Web functionality was rated of most importance to users. However, user perception of library success in this area was barely above the established minimum level.[7]

What's needed for "personal control" to succeed are new digital access and management technologies. Yet, to state that any new digital technology offers challenges and opportunities to libraries is perhaps, the most pervasive current cliché that can be offered. Nevertheless it remains true, if only because there does not seem to be any technology that does not shackle as it liberates. One example of a promising technology is the portal or gateway. A portal or gateway can be a simple collection of Internet resources conveniently collected on a single Web page and allowing users to begin search or exploration. This is, for many users, a home on the Web, a start and return point where information can be stored and communication begun. It is, in a precise coinage, a pocketweb. More importantly, however, portals also allow customization and personalization options so that the most relevant resources from a default set can be added while those deemed less important will not appear. In a portal environment such as MyYahoo or MyNetscape making these choices is relatively unproblematic. Is your sports priority the NBA or the NFL, or something else? Do you want weather forecasts from your city or another city? Stock quotes from NYSE or NASDAQ or a combination of the two? Do you want a thought of the day, a horoscope, a recipe? Click here. No matter the number of these choices, they require only decisions based on everyday life. Beyond the addition of a Web search engine, personal Web bookmarks and some optional layout choices such as color, font, and page design these portals most closely resemble a customized newspaper.

There are more complex variants on the portal idea. In the business world, the Enterprise Information Portal has recently been touted as the next "killer application." EIPs are designed to provide a much more complex collocation of function than a simple Web portal such as MyYahoo. A typical definition of an EIP from the portal vendor Hummingbird is replete with an impressive assemblage of terms:

> EIPs provide a fully customizable Web-based environment to access the Internet, Intranet and Extranet, click stream data, mission-critical legacy applications, enterprise business packages, customer relationship management (CRM) solutions, custom applications, custom business solutions, and other vital applications. EIPs allow users personalized, seamless integration of content and commerce, and the ability to relate with others through communities.[8]

What Hummingbird is essentially defining here is that a truly effective enterprise information portal solution must include, as a base point, the following:

- A single point of access (single login)
- Unified search across all information sources
- Personalization
- Information and application integration
- Categorization
- Collaboration
- System security
- Scalability
- Openness

Such functionality is common in different definitions. The uPortal project defines a portal as: "A one-stop client-oriented web site that personalizes the portal's tools and information to the specific needs and characteristics of the person visiting the site, using information from university databases." The Resource Discovery Network subject portals project from the United Kingdom's Joint Information Systems Committee lists basic requirements as:

- Customisable home pages driven by a single secure log-in
- Ability to share information and communicate across the community
- Transparent access to a wide range of high-quality information and information services that have been deemed to be of relevance in a particular context
- Ease of use and ubiquitous availability
- Access to information located in disparate data sources, including unstructured data.[9]

None of these elements are controversial, whether the portal is being built for a Fortune 500 company, a university campus, or a library. What all portals must provide is the ability to sample all relevant information without consuming all relevant information every time specific information is required. In other words all portals must protect against information overload. In a rapidly changing and emerging digital environment such a requirement seems trivially obvious. Nevertheless such a requirement has some operational difficulties to overcome before success can be confirmed. In the matter of library portals the primary difficulty is deciding how to negotiate the customization and personalization options from the librarian's point of view (i.e., this professional community of discourse) to that of, as a base point, the average undergraduate. What is required is an amalgamation of the outreach activities of a public service reference desk and the structure and access points of an online public access catalog. Certain social factors make this more difficult than simply using previous practice. Of primary importance is that this is an unmediated experience. Personalizing and customizing of a library portal most often takes place online. Consequently, the process must be transparently obvious. In a more specific sense, the choices made are not from everyday experience. We do not assume, of course, the possibility that students have internalized the idea of the research library to their everyday experience. This might be a vain hope but a hope that nevertheless springs eternal in the hearts of librarians.

A library portal can be positioned to exist as a virtual analog of some of the basic features of a reference desk. As the reference desk can be the first place to gain an overview of the resources most pertinent to a research question so too can a library portal, once properly customized, serve the same function. A functional library portal can display such variables as the most relevant databases and electronic journals in a subject area, user-selected Web bookmarks, library recommended Internet resources, lists of new titles in selected call number ranges, and other similar resources.

In order for a portal to be more than a flat file Web page, the page must be customized to the research interests of the user beginning to use the service. And in order for this to occur there must be some manner of registration dialog or profile that allows the new user to state certain preferences. It is the initial interaction between user and system that defines success or failure. Many factors contribute to this success of this interaction but perhaps no element is more important than context. As Borgmann notes context is one of the five factors needed to produce information.

A recent article on enterprise information portals discusses a business portal requirement that can be readily applied to library portals. In order to derive business benefits from an EIP a common level of understanding is required and for this to occur the portal and its underlying infrastructure must support context. If a portal is a context provider then it must provide context through logical constructs that can be thought of as a corporate semantic. The corporate semantic provides the common understanding that elicits the proper information.[10] Unlike a corporate semantic, however, a library semantic must provide a common understanding for the users of the portal and not just for librarians. In a library portal, context is provided through a library semantic that provides the necessary information so that meaning can be developed from the tools at hand. The question is how do libraries develop a semantic that adequately explains both the analog and digital information environments. In order to set the stage for that portion of our discussion we need to briefly examine certain communication processes in the university library.

If we were to define a library semantic, surely classification schemata would be the bedrock of such a language. For, the essence of cataloging, or more properly, of bibliographic control, is to enable a person to find any intellectual creation, whether issued in print or nonprint format, when a variable such as author, title, subject or other element is known. The precise description of objects such as a monograph, journal, report, or CD-ROM aids in the retrieval and evaluation of such objects. That is why over the years libraries have developed specific vocabularies that allow such description. Without such classification vocabularies description becomes diffuse, obscure, and any resultant retrieval less accurate, if not impossible. A precise vocabulary brings in its train a necessary complexity. This is not complexity for its own sake or for the sake of the cataloger. In 1915 University of Michigan Librarian William Bishop observed that "catalogs are complex because people and books are complex."[11] What was true nearly ninety years ago is no less true in today's digital environment. John Seely Brown recently made much the same point when he said:

> On the Web, most information does not have an institutional warranty behind it, which really means you have to exercise much more judgment . . . . If you find something in a library, you do not have to think very hard about its believability. If you find it on the Web, you have to think pretty hard.[12]

That something is true, however, does not make it obvious. What is obvious is that bibliographic control has slammed full speed into the brave new world of random access. As Eric Ormsby wryly notes: "Once it became feasible to conduct Boolean searches or simply to employ keywords, the entire edifice of subject classification and taxonomy began to totter." This interconnectedness of everything means "anyone can now find *something* in the library as long as he has even the dimmest idea of what he is looking for."[13]

Such observations are now neither new nor rare. A reason for their ubiquity is the professional recognition that the ever-increasing influx of digital materials will increase the amount of material that can be described and indexed and accessed without human intervention. The question is how much of this is a good thing for library patrons, if not for librarians? The result of patron dependence on digital materials is obvious. Clifford Lynch notes one particular effect:

> Destiny may be digital, but we will be a long time reaching this destiny, and this long transitional period will call for careful management. We are already seeing print collections in our great libraries beginning to fade into invisibility for many patrons; materials available in digital form are so conveniently available, and so much more accessible . . . that for these patrons the collection may as well only contain the digital content.[14]

As specialization has increased librarians have also developed specialized vocabularies to talk to each other. Like many professions these are often derided from both within and without as mere jargon. Nevertheless such vocabularies play a valuable role in fostering accurate and timely discussion. In library departments from collection management, to public services, to cataloging, to access and delivery, librarians can rely on the fact that colleagues share a general understanding about their professional community. This is not to deny, however, that occasional inter-departmental rivalries can delay complete equanimity on all matters.

It might be fairly argued that noting the complexity of either bibliographic control or of the professional vocabularies of librarians does not make it impossible to explain to the average undergraduate how to select and access appropriate information. Is this not what a public service desk has always been doing? Reference librarians function as mediators between the complexities of a sophisticated information environment and students having no profound understanding of that environment and yet an equally urgent need to find appropriate information. As reference librarians have long noted, there are two problems to be overcome. First, students often do not possess a clear understanding of what they want. This is often referred to as the problem of the "real question." These questions are often further obscured by the student's own use of jargon derived from imitating their professors but with an incomplete understanding of the ideas originally expressed.

The second problem is that of user expectation. Ormsby observes that "over the last two decades . . . research in libraries, among graduate and undergraduate students, is getting shallower and shallower."[15] Despite a deluge of publicity, a "facility in surfing the internet is no substitute for the struggle to understand." Put another way, if we assume that good metadata expedites quality research and if the Internet is rife with sloppy metadata, then students, in their reliance on Google, cannot be expected to produce quality research. While variants of these criticisms have existed for some time, it is an open question if this is not now an increasingly serious obstacle. If there are indeed significant gaps between user and librarian, student and teacher, student and student, and even librarian and librarian, a complex digital information environment only makes it worse (or challenging, for the more optimistic).

In 1993 George Ritzer, drawing on Max Weber's theories on the effects of mass bureaucratization, argued that modern society is exhibiting signs of what he termed "McDonaldization." "McDonaldization is the process by which the principles of the fast-food restaurant are coming to dominate more and more sectors of American society as well as the rest of the world."[16] Ritzer examines the structure and process of McDonalds and finds five dominant themes: efficiency, calculability, predictability, increased control, and the replacement of human by non-human technology. We can find many examples by focusing on just the theme of efficiency. Consider the drive-up window, salad bars, fill your own cup, bus your own table, self-serve gasoline, ATMs, Voice Mail, microwave dinners and supermarkets. The interesting element here is that the customer often ends up doing the work that previously was done for them. People end up spending more time, being forced to learn new technologies, remember more numbers, and often pay higher prices in order for the business to operate more efficiently (and hence maintain a higher profit margin). Ritzer's ideas have been both influential and controversial. There were those who thought he went too far and those who thought he did not go far enough. Nevertheless the idea is provocative and was recently applied to the university library. Brian Quinn noted "Students themselves may contribute to the McDonaldization process by approaching the university and the library as consumers would."[17] He sees the library environment in similar terms:

> Students want short lines, polite and efficient personnel, and the flexibility to 'have it their way.' . . . many students find computers faster and easier to use than paper sources . . . The quality of the information becomes secondary. In turn, librarians must acquiesce by providing the 'Information Happy Meals' the students are seeking in order to guarantee 'customer satisfaction.'

> Many academic librarians have a goal to help create independent lifelong learners, but some students regard library research as being too

much like work. The result is a 'dumbing down' of reference services in order to placate the student. In the McDonaldized library, 'the customer is king,' which essentially means giving students what they want rather than what they need.[18]

It is not the purpose of this paper to assert the correctness of either Ritzer or Quinn's interpretations; however, such ideas serve as provocations to think about the implementation factors inherent in customizing and personalizing a library portal. The connection between these social observations and the library portal begin in the acknowledgement that technology does possess the power to make things easier. This is perhaps the coeval virtue and vice of technology, it is often easier to "dumb things down" than to smarten them up. And since making things easier through uniformity and predictability in the name of efficiency is an omnipresent factor, then, in this Ritzer is surely correct.

The purpose of a library portal is, in part, an effort to make the research process more transparent and integrated into the teaching and learning environment. It can organize and present research materials in an efficient manner, but it cannot make things easier. The library Web portal format, with its attendant customization options, encourages comparison with the commercial Web environment. Nevertheless, it is critical to remember that a retail or entertainment Web site is inherently directed at the individual and inherently motivated to make it easy as possible to use. There is in a retail site such as Land's End or L.L. Bean an essential sameness of material–socks are socks, pants are pants. Apples are never oranges. While one might argue that a bibliographic citation is a bibliographic citation and an electronic journal is an electronic journal, the mere possession of such material is but one step in a complex process of research and discovery. We must continually bear in mind that research is not a push button process. Social pressure might coerce reference librarians into supplying users with the required number of citations, nicely printed and formatted but the disservice is in promoting the assumption that the process is easy.

From this overview of portals and an exploration of Ritzer's thesis, let us return to the thread concerning language and communication that began with those Inuksuit standing sentinel on the treeless tundra. To paraphrase Dickens–with extreme liberality, for which we beg pardon–language thwarts communication, and this must be distinctly understood or none of what follows will seem useful. Implementation of a MyLibrary service runs into this axiom again and again, so it is best to be clear on the point. To do this, we turn from paraphrase to quotations, this time of philosopher Frank Smith:

> Natural language (the languages we all learned so effortlessly in childhood) are not very good for purposes of communication. They are not

very effective for the unambiguous communication of information, knowledge, or data. . . . The reason language is so ineffective for conveying information is its ambiguity and lack of direct correspondence with the world. . . . Language always needs to be interpreted; it is never explicit.[19]

A portal service is a tool for the user, a way for her to extend her reach into digital realms and, as swiftly and with as little cost to her finite store of attention as possible, bring back such stuff as knowledge can be made of. Ideally, this means MyLibrary becomes her home page, and if using it is difficult of mystifying,[20] it will drop from use faster than Enron stock plummeted down the Exchange.

One way to make the service mystifying is to use library jargon and acronyms: OPAC, CAM, LC. This excludes the user, causing him to wander off in search of some other provider who is willing to put forth effort in pursuit of clear communication. The solution is to replace such terms where possible and provide brief explanatory text otherwise. At NC State, OPAC became *catalog*, CAM–the Current Awareness Management feature of MyLibrary–became *New Titles*, and LC became *call numbers*, this last hyperlinked to a brief schedule of the Library of Congress classification system.

Usability testing can help in the fight against mystifying professional cant: in MyLibrary@NCState, a proposal to change the *University Links* category to *Community Resources* foundered on the shoals of just such a test. MyLibrary surveys go through testing before release and these tests turn up enough unfounded assumptions about what users know to more than earn their keep. As a result of careful scrutiny, user-friendly constructions such as "Suggest A Purchase" and "Request Items" now stud a typical MyLibrary Page. NC State has attempted to speak fluent "user-ese," but work yet remains: a service by the evocative yet less than transparent name of "TripSaver" still haunts the roster of the Library Links category.

An aspect of MyLibrary that is perhaps beyond the means of usability testing to fix, because it involves factors outside the library's control, is the interaction between the user and the controlled vocabulary of discipline names, the spine of the NC State Libraries portal service. At least four main taxonomies of knowledge exist at NC State, five if one counts the Library of Congress classification schedule: MyLibrary's discipline list, the NCSU Libraries Website, the university's list of academic departments, and the schedule of courses. The library controls only the first two of these, and even those have proven difficult to align. This profusion of vocabularies has generated as much discussion regarding MyLibrary@NCState, perhaps more, than any other subject.

For instance, users involved in Family & Consumer Sciences, 4-H Youth Development, Physical Education, or the three ROTC programs, all academic departments at NCSU, will find no corresponding discipline in MyLibrary, while Psychology can expect a one-to-one fit and those profiled in Physics can at least find *Physics and Astronomy*. And how does an undergraduate, enrolled in an Ecology course, find appropriate resources in MyLibrary? Does she try the *Natural Resources* discipline? Sounds as if that might work, but what about *Environmental Science*, or perhaps *Environmental Sustainability*? Since they all seem to have some relation to Ecology, which offers the best fit? Are tools provided to help her choose?

Should anyone think all this amounts to a tempest in a teapot, consider a recent message to the MyLibrary listserv: a person was disappointed not to find his discipline, Landscape Architecture, on MyLibrary's list, and pointed out that it is quite different from Architecture, which in MyLibrary forms one-third of an *Art, Architecture, and Design* troika. He even offered to help select the resources critical for his discipline. Language and intentionality are here thrown into relief. For MyLibrary, disciplines are intended to serve as a way to create relationships between resources and users, librarians and resources, and librarians and users in the database that undergirds the service, but for users, there are important issues of identity wrapped up here, and they play out among these discipline lists, as if inclusion by the library constitutes validation of a career.

Some of the examples mentioned above bear revisiting. In general, MyLibrary@NCState contains specific disciplines, such as *Zoology* and *Music*, but some, such as the aforementioned *Art, Architecture, and Design*, which includes several departments, create problems for content providers when they try to select default resources. However, alleviating this problem by aligning MyLibrary's discipline list with the list of academic departments–which translates to adding fields of study to the present menu of 67–would only create more work for the content providers: someone would have to maintain those new disciplines.

A note about maintenance of the service. Should one implement a MyLibrary-style portal, no other issue will be so charged with politics, here defined as garnering staff buy-in. The workload placed on content providers by MyLibrary is enormous. Here at NC State they select default resources for disciplines in their areas–plural because most have more than one–keep up with new resources licensed by the library and those discontinued, guard against link rot, that affliction that results in 404 messages, scout for Internet resources to add and then tag them for use in MyLibrary using CORC (Cooperative Online Resource Catalog), and regularly use the *Messages from the Librarian* feature. If content providers sense there is less than full support for this service on the part of administration, support that manifests itself in the form of either hiring new staff in recognition of the fact that there will be more work, not less, or by reducing the expectations and responsibilities of the current staff, the portal service will never get off the ground. A word to the wise, there.

Now that politics has been mentioned, let us return to *Environmental Sustainability*. It, along with three other interdisciplinary areas of study that appeared in a recent compact plan by NCSU, was included in the MyLibrary discipline list as a way to anticipate such a request from university administration, tacit acknowledgment that MyLibrary@NCState exists within the context of a research university and that such a context is not without effect. The amount of intellectual terrain spanned by these areas is vast; not only did the decision play hob with attempts to pick default resources but it also caused discussion to turn to slippery slopes: where to draw the line on interdisciplinary areas? What happens to the integrity of the discipline list when such are added? Did inclusion of these high-profile areas in MyLibrary help to validate them, or was it the opposite–an attempt to validate MyLibrary as an important player in the university scene? And above all, what do users think of such categories? Will they be confused, seeing four such broad areas included among what are, for the most part, single disciplines, or will they care? Though it may amount to telling secrets out of school, MyLibrary@NCState has not taken the campus by storm–it seems more famous abroad than it does here at home.

But these are fixed features of the service, names of categories that appear on a MyLibrary page or are discovered in a drop-down menu. They change but little, for the good reason that users need to have faith that the discipline in which they have profiled themselves will not suddenly vanish. This is probably the best reason why a discipline vocabulary should be selected with care. Problems in these fixed features once rooted out and corrected, cease plaguing the user.

Trickier is the interactive component, that rectangle of real estate on a MyLibrary page variously named Message of the Day or Message from my Librarian. This is where the librarian pens timely, concise notes to the user, messages concerning new resources added in the user's discipline, upcoming talks and seminars that may be of interest, and so forth. Easy for librarian patois to slip in here. Even a locution as apparently innocuous as "literature search" may cause problems, particularly for an undergraduate. At the least, this means it pays to know one's audience and think carefully about how one addresses it.

This leads to more bad news from Smith, because MyLibrary's interactivity is powered by written language, which "is even more opaque than speech, despite its undeserved reputation for lucidity. Written language lies wantonly on the page, oozing equivocation."[21] Or on the screen, in this case. Messages from the Librarian are kept brief not just to save precious screen space but also because, "Reducing a statement to a few . . . sentences reduces disputation because there is less to argue about."[22] Less chance of inadvertently falling into library cant, as it were, and here we pause for another axiom, courtesy of Karl Popper: precision in language comes only at the cost of clarity, or, if that one page memo confused employees, doubling its length is unlikely to do anything but double the confusion. We are never as perspicuous as we would like to believe. It is a grim credo, but it saves much long-term anguish.

To take the sting out of what has just been said, remember what MyLibrary is trying to accomplish–serving not just as a self-service research tool but also, somehow, as a virtual analog for the reference desk. In a successful reference interaction, the librarian decodes a need expressed by the user and then interprets for that user the complexity of the collection. A difficult job, but in MyLibrary this process must be digitally replicated, i.e., without the suite of nonverbal cues that give spoken language an edge over written when it comes to communication. Faced with such a task, it is helpful to have no illusion regarding the supposed lucidity of written language.

One way to compensate for this quixotic challenge is to cede control over the front-end to the user. They should be able to rearrange categories on their page, rename them, show some and hide those they have no use for, and combine resources in categories of their own devising. In its current state, the classification of resources used by NC State librarians on the back-end–indexes and abstracts, e-journals reference shelf–also determines placement of same on the front end. How much better if a user could take selected e-journals, shuffle in a few indexes, spice with an online subject encyclopedia and the Website of a professional organization, and call the resultant concoction whatever she pleases: English 111, if she's an undergraduate, or Dissertation Resources if a Ph.D. candidate. To enable this change would require completely overhauling the tables that compose MyLibrary's database, a daunting task, and yet a necessary one if MyLibrary is to be successfully marketed to a university community, to say nothing of competing with commercial portals for a user's attention. The point is, MyLibrary resources can be as complexly classified on the back-end as one wishes to make them, but none of that complexity should appear on the user's computer screen.

The Inuit are able to use the Inukshuk because they share a common semantic. To the untrained eye they are but piles of stone. As we have illustrated, they are instead conveyors of information. The challenge to the librarian community is to construct a semantic that speaks to a user community continually bombarded with the message that faster is better, quantity trumps quality, and that superficial flash defines excellence. But we must ask ourselves, is this the best way to do that "research stuff"?

REFERENCES

1. This very simplified description of the Inuksuit owes much to Norman Hallenday's *Inuksuit: Silent Messengers Of The Arctic* (Douglas & McIntyre/University of Washington Press, 2000). For a more sophisticated explanation of the differing types of Inuksuit see pp. 46-47, 116-118. For illustrations of various Inuksuit cf. <http://www.aventurecanada. com/images/paysage/grand-nord/village_inukshuk.jpg> (Accessed December 3, 2001) and <http://www.canada.worldweb.com/PhotoGallery/LandmarksandHistoSites/10-2079.html> (Accessed December 3, 2001).

2. Hallenday, p. 46.

3. Albert Borgmann, *Holding On To Reality: The Nature of Information at the Turn of the Millennium* (Chicago: University of Chicago Press, 1999), 22.

4. Borgmann, p. 213.

5. Warren Thorngate. "On Paying Attention," in *Recent Trends in Theoretical Psychology*, edited by William J. Baker et al. (New York: Springer-Verlag, 1988), 247-263.

6. David Levy, *Scrolling Forward: Making Sense of Documents in the Digital Age* (New York: Arcade Publishing, 2001), 113.

7. Colleen Cook. "TRLN Symposium: Introducing LibQUAL+." *http://www.lib. ncsu.edu/staff/kamorgan/libqual/cook.htm* See slide 28. Further details of the ARL LibQUAL+ survey are at http://www.arl.org/libqual/ (Accessed December 7, 2001).

8. Hummingbird EIP datasheet. For more information on the elements of an enterprise information portal see the Delphi White Paper "An Enterprise Portal to E-Business" at <http://www.hummingbird.com> <http://www.hummingbird.com/products/eip/index.html> (All accessed December 3, 2001).

9. More information on uPortal is at <http://mis105.mis.udel.edu/ja-sig/uportal>. The JISC subject portals are discussed at <http://www.ariadne.ac.uk/issue29/clark/>. An interesting discussion on elements of a campus portal is accessible at <http://www.cren.net/know/techtalk/events/portals.html> (All accessed December 3, 2001).

10. Robert Bolds, "Enterprise Information Portals: Portals in Puberty," *Special Supplement to KMWorld* (July/August 2001): S16. <http://www.kmworld.com/publications/whitepapers/portals/bolds.htm> (Accessed December 3, 2001).

11. William Warner Bishop. *The Backs of Books and Other Essays in Librarianship* (Baltimore, Williams and Wilkins, 1926), 134.

12. John Seely Brown, as quoted in Sarah Thomas', "The Catalog as Portal to the Internet" <http://lcweb.loc.gov/catdir/bibcontrol/thomas_paper.html> (Accessed December 8, 2001).

13. Eric Ormsby, "The Battle of the Book: The Research Library Today," *The New Criterion* 20, no. 2(October, 2001): 9.

14. Clifford Lynch, "The New Context for Bibliographic Control in the New Millennium" <http://lcweb.loc.gov/catdir/bibcontrol/lynch_paper.html> (Accessed December 3, 2001).

15. Ormsby.

16. George Ritzer, *The McDonaldization of Society: An Investigation Into the Changing Character of Contemporary Social Life* (Thousand Oaks, Calif.: Pine Forge Press, 1993), 1.

17. Brian Quinn, "The McDonaldization of Academic Libraries?" *College & Research Libraries* 61, no. 3 (May 2000): 248-61.

18. Quinn, 249.

19. Frank Smith, *To Think*. (New York: Teachers College Press, 1990), 110-111.

20. Although this discussion focuses on the general characteristics of library portals, it is, necessarily, informed by the author's experience with MyLibrary@NCState. Further information on the NCSU Libraries portal is available in our "Pioneering Portals: MyLibrary@NCState," *Information Technology and Libraries*, 19:4 (December 2000): 191-198. Accessible also at <http://www.lib.ncsu.edu/staff/kamorgan/pioneer. html>.

21. Smith, 111.

22. Smith, 112.

HILT:
Moving Towards Interoperability
in Subject Terminologies

Dennis Nicholson
Gordon Dunsire
Susannah Neill

SUMMARY. The HILT ('HIgh-Level Thesaurus') project was a UK based and focused desk-study of the problems associated with cross-searching and browsing by subject in a cross-sectoral and cross-domain environment encompassing libraries, archives, museums, and electronic resource collections. It aimed to reach a community consensus on the best way of at-

Dennis Nicholson is Director of Research, Information Resources Directorate, Strathclyde University and Director of the Centre for Digital Library Research. He was Director of HILT and has managed a range of other projects including BUBL, the CATRIONA project, CAIRNS distributed catalogue project, and others (for a full list of current CDLR projects see the CDLR projects page at: http://cdlr.strath.ac.uk/projects.htm).

Gordon Dunsire is Research and Projects Manager, Napier University Learning Information Resources. He is Chair of Cataloguing and Indexing Group in Scotland and played a key role in the formulation of the CAIRNS cataloguing and indexing standards. He is a member of a Consortium of European Research Libraries (CERL) Manuscript Working Party investigating the feasibility of using Z39.50 to create a distributed manuscripts catalogue in Europe.

Susannah Neill is New Technologies Development Officer, Department for Lifelong Learning, University of Wales, Bangor. She was Research Assistant on the HILT Project based in the Centre for Digital Library Research.

[Haworth co-indexing entry note]: "HILT: Moving Towards Interoperability in Subject Terminologies." Nicholson, Dennis, Gordon Dunsire, and Susannah Neill. Co-published simultaneously in *Journal of Internet Cataloging* (The Haworth Information Press, an imprint of The Haworth Press, Inc.) Vol. 5, No. 4, 2002, pp. 97-111; and: *High-Level Subject Access Tools and Techniques in Internet Cataloging* (ed: Judith R. Ahronheim) The Haworth Information Press, an imprint of The Haworth Press, Inc., 2002, pp. 97-111. Single or multiple copies of this article are available for a fee from The Haworth Document Delivery Service [1-800-HAWORTH, 9:00 a.m. - 5:00 p.m. (EST). E-mail address: getinfo@haworthpressinc.com].

tempting to solve interoperability problems arising out of the use of different subject terminologies by services in these areas and obtained wide support across the communities for the proposed way forward–a pilot terminologies mapping service and associated participatory process aimed at supporting and facilitating the maintenance of a consensual approach to ongoing development. *[Article copies available for a fee from The Haworth Document Delivery Service: 1-800-HAWORTH. E-mail address: <getinfo@ haworthpressinc.com> Website: <http://www.HaworthPress.com> © 2002 by The Haworth Press, Inc. All rights reserved.]*

KEYWORDS. HILT, thesauri, mapping, subject search, subject browse, interoperability

INTRODUCTION

The problems of semantic interoperability between different vocabularies and categorisation schemes have been understood by library classifiers for many years, but there was little perceived need to co-ordinate or standardise until the evolution of technologies for connecting many different cataloguers to non-local enquirers. The issue has now, in a short space of time, become acute, and is adversely affecting the benefits of resource sharing, collection management, and information retrieval which should accrue from wide-area networking.[1]

The HILT[2] ('HIgh-Level Thesaurus') project was a UK based and focused desk-study of the problems associated with cross-searching and browsing by subject in a cross-sectoral and cross-domain environment encompassing libraries, archives, museums, and electronic resource collections. A collaborative project involving a number of cross-domain participants,[3] it aimed to determine whether there were likely to be interoperability problems arising out of the use of different subject terminologies by services in these areas and to propose a way of solving any problems identified. A key requirement from the project funders was to determine whether it was possible to reach a consensus across the communities in respect of a way forward and to base the project's recommendations on any consensus reached. Having determined that many different subject schemes and practices were in use in the UK services surveyed, the project was subsequently able to report a strong consensus across the communities on the best response to this and other aspects of the situation uncovered.[4] The favoured approach was to move to a more practically focused follow-up pilot project that would develop, and accurately determine the full costs and benefits of, "a networked, user and machine responsive, interactive route map to the terminologies used by the communities and the relationships between these," an outcome that echoed the project's own findings. Appendix A presents a diagrammatic description of what is meant by a terminologies route map (or 'TeRM') service. At the time of writing, the final report of the project is close to completion, but not entirely finalised. However, any future changes to the detail of this or indeed the report itself are likely to be minor.

OF TERMS AND TeRMs

A TeRM based on existing commercial software would

- interact with users to clarify the context of subject terms input; e.g., 'lotus': the flower, or the software, or the car, etc.;
- 'output' a pre-configurable boolean search combining appropriate words from a range of schemes and relationships specified by the service provider;
- then permit the resulting search to be used to search indexes of internet resources, either 'as is' or after subsequent amendment by the user.[5]

Work carried out previously in the context of the CAIRNS[6] Z39.50-based distributed catalogue project suggests that the software could be developed further to allow both the searching of such distributed catalogues and the landscaping required to narrow down the range of services to a subset relevant to the user's requirements, e.g., only archives, only museums, only libraries in Scotland. Under consideration are schemes like LCSH, DDC, AAT, and UNESCO, along with UK adaptations and extensions of these implemented in key UK services such as the national libraries and equivalents or near equivalents in the archives, museums, and electronic services domains, but any set of schemes could be utilised. The suggestion made is that any development take place in an international context, with a UK TeRM aiming to be a model for others elsewhere. This would reduce the overall cost of set-up and maintenance, amongst other things, provided the right co-operative funding model could be agreed and implemented.

FIRST INDICATIONS–FOCUS GROUP OUTCOMES

The first indication that this approach might provide the best basis for a consensus came from a small Focus Group, set up early in the project with a view to informing the HILT team about issues and attitudes in the various domains.

The group comprised one representative from each of the archives, libraries, museums and online information services domains. They were asked to discuss a series of issues drawn up by the HILT team in order to gain a greater understanding of the differences and similarities between the sectors and therefore some of the problems that might be faced. A number of useful insights were gained, including an early indication that:

• Consensus based on a single subject scheme was unlikely, given the problems raised–and views expressed–by participants.
• Mapping or switching between schemes might provide such a basis, but there was a perceived need to investigate the terminology used by users themselves, and to create user friendly front-end search tools which could accommodate the complexities of any schemes mapped.[7]

CONSENSUS–THE STAKEHOLDER WORKSHOP

Confirmation of these early indications came much later in the project when a Stakeholder Workshop was held with the specific aim of trying to reach a consensus on the best way forward in respect of the situation uncovered by the HILT team. The workshop brought together a balanced and representative set of around 50 participants from archives, libraries, museums, and online information services. Sessions before lunch were informative. They consisted of a number of presentations given by stakeholders, partners and consultants on the different perspectives of the various communities to the issues HILT was addressing, along with others on the costs and problems of mapping, thesaurus interface issues, and, with the future in mind, the semantic Web. These were followed by an afternoon session in which four pre-designed Breakout Groups, balanced to ensure a fair representation of each community, were given the task of reaching a common view on a number of issues outlined in the document 'HILT Workshop Breakout Sessions: Discussion Issues (and notes).'[8]

In the context of these sessions, the delegates, who had all been given details of the options and background information prior to the day, were asked to attempt to reach a consensus on the basis of one or more, or some combination of, the following set of options:[9]

[OS1] Do Nothing

1. Artificial Intelligence will solve it in time
2. Big business–Microsoft or similar–will solve it

3. It's not really that important
4. No solution is necessary
5. The problem cannot be solved

[OS2] Set Up a Human Process Intended to Lead to a Solution in Time

1. Set up a Terminologies Agency, perhaps based on National Libraries/mda/NCA
2. Set up an inter-domain, inter-sectoral Task Force to move the communities towards a solution
3. Set up a Terminologies Agency and a Task Force

[Note: A description of what is meant by a "UK Terminologies Agency" was also provided.]

[OS3] Adopt a Base-Level, Gradual Approach,
with an Eye on Future Developments

1. Adopt a single scheme such as DDC and apply to all collection level descriptions in the UK
2. Gradually create inter-service and inter-community terminology 'cross-walks' eventually building up to a partial but adequately broad solution
3. Aim to solve the problem for electronic services only, perhaps via the semantic Web vision
4. Provide more flexible retrieval facilities for users
5. One or more of these four together (please specify)

[OS4] Adopt a Single Scheme

1. Adopt: LCSH/UNESCO/DDC/UDC/AAT/Another scheme (say which)/A New Scheme in addition to the existing scheme used by any given service [Please specify which scheme]
2. Adopt: LCSH/UNESCO/DDC/UDC/AAT/Another scheme (say which)/A New Scheme instead of the existing scheme used by any given service [Please specify which scheme]
3. Adopt a single scheme: without retroconversion of legacy metadata/ with retroconversion funded by the host organisation/with retroconversion funded centrally [Specify which]

[OS5] Mapping Service Alternatives

1. Set up a mapping service, ideally with international participation and support, and gradually build towards a complete mapping of LCSH, UNESCO, UDC, and AAT to a DDC backbone. Include local adapta-

tions and extensions from major services such as the National Libraries. Use the international service with the mapping of UK adaptations and extensions as a model for other countries. Determine and implement the best international funding and maintenance model.
2. Set up a 2 year mapping service pilot to measure costs against benefits of both a full-scale service and all of the various alternative responses detailed on this page.

[Note: A description of a mapping service was also provided.]

The outcome was strongly in favour of the kind of approach that the project subsequently labelled an interactive terminologies route map or TeRM service. All four Breakout Groups agreed that Option Set 5–the mapping service–was the preferred option set. In groups A and D this was a unanimous decision. Group B reported that 8 out of 9 members favoured it and Group C answered that 'a majority' chose it. Option 5.2–a 2 year pilot mapping service–was chosen by all of the four groups as the preferred option *within* option set 5. Results from the groups were unanimous for 2 groups, and 8 out of 11 and 7 out of 9 for the others. Group C was unanimously in favour of combining option 5.2 with the option of establishing a task force to look at user needs, the costs of rolling out and sustaining a mapping agency, and the cost-benefit analysis of a sustainable solution.

THE INTERIM REPORT–A FINAL CONFIRMATION

In order to enhance the chances of a clear outcome at the workshop, amongst other things, delegates were promised that they would have a further chance to comment on the results by being asked to react to the subsequent Interim Report, the contents of which were largely based on the workshop and its outcomes.[10] This consultation exercise achieved an 86% return rate. Forty-eight responses were returned out of a total of 56. Non-responses were believed to be due to:

• Abstention, being unable to reply on behalf of their organisation.
• Holidays (unfortunately, the exercise had to be conducted over July and August).

In addition, 5 respondents (11%) gave personal rather than organisational opinions, either because they were unable to reply on behalf of their organisation but wanted to express an opinion, or because they represented themselves, e.g., they were consultants. The results in summary were:

- 21 respondents–45%–agree.
- 20 respondents–43%–agree, with reservations.
- 3 respondents–6%–don't know.
- 2 respondents–4%–disagree, with reservations.
- 1 respondent–2%–disagree.

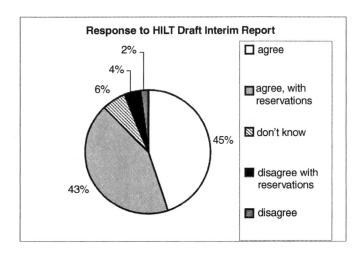

Overall, 88% of respondents endorsed HILT's interim recommendations in favour of a pilot TeRM, as it was later called, with or without reservations; results within the archives, libraries, museums and heritage, and online information services sectors were broadly similar. In addition, a number of issues were raised under the heading of 'reservations' which were subsequently built into HILT's full set of final recommendations.

PROBLEMS AND THE NEED FOR A PILOT

The proposed approach has several features to recommend it. First, there are a number of unknowns: questions that cannot be answered definitively without the programme of empirical tests envisaged within the proposed pilot follow-up project. Further research is required into the level and nature of user needs, the effectiveness of a TeRM service in meeting them, and the practicality, design requirements, and costs against benefits of such a service when compared with other possible solutions, before a long-term commitment to a possibly expensive enterprise could be justified. This is best done via a pilot project based on an interactive route map of the type described; it is the only one of the various possible solutions to the problem of ensuring interoperability in respect of subject terminologies that permits all of the possible solutions to be tested without requiring what would be major but necessarily temporary changes to service databases.

The following extracts from the report of the HILT literature highlight two of the key areas of concern and give some indication of the elements to be considered.

Users

There is very little information available in the literature on the needs and behaviour of users as regards subject searching in a distributed environment, particularly in respect of users associated with the UK-wide, cross-domain environment being considered by HILT. Relevant papers include Pollitt's[11] reviews of developments at the Universities of Huddersfield and Leeds regarding end user interfaces for online database searching, Gillis's[12] consideration of issues involved in 'sub-clump' searching on CAIRNS, utilised to narrow cross-searching by subject area, and Burnett and McClure's[13] description of the preparation for the new Museum of Scotland where the focus was on user support and education and no attempt was made to facilitate cross-searching amongst relevant databases.

A recurrent theme in many papers was a perceived need for services or service staff to provide assistance to users searching or cross-searching by subject, a point generally agreed on by HILT participants, and subsequently regarded as an essential aspect of the HILT recommendations. Jascó,[14] for example, states that, in many cases, databases do not offer the necessary aids required to use the "preferred terms of the subject controlled vocabulary." The databases to which Jascó refers are Dialog and DataStar. However, the situation he describes may be viewed as having some bearing on HILT. Others look at the presence, rather than the absence of such aids:

- Hug[15] looks at subject access to the catalogue of the Swiss Federal Institute of Technology in Zurich (ETH-Bibliothek) where entries are classified with UDC numbers, subject access is based on a subject register linking terms in German, English and French to UDC numbers, and the user doesn't need to know UDC numbers to search the catalogue.
- Doszkocs[16] examines a system developed at the U.S. National Library of Medicine for automatically mapping users' natural language search

terms to a controlled vocabulary, in this case MeSH. The user interface then provides potential search terms from MeSH and from text and abstracts of documents.

- Schatz et al.[17] report on a prototype user interface developed at the University of Illinois Digital Library Initiative (DLI). This supports the principle of multiple views, where different kinds of term suggestors can complement the search and each other and describes the design of the retrieval system. However, no experimental results are given on use.
- Cochrane[18] describes the reasons behind, and the methods of developing, a design for DDC which allows for the display of the schedules and indexes in a hypertextual browser which was originally designed to display conventional thesauri. A summary of previous research into the search strategies and problems of OPAC users is given. The intended use is that library users would be able to browse through and explore Dewey before choosing their search terms or numbers.

Time, Cost and Effort in Mapping, Integrating and Creating Controlled Vocabularies

The majority of articles found in the literature go into very little detail on the time, cost and effort involved in mapping controlled vocabularies. What detailed information the literature survey did produce was project or service specific.

However, Aitchison et al.[19] give some general advice on resources required for thesaurus construction, which could also be applied to mapping and integration:

- *Financial resources:* Depends greatly on the size of group/institution/service. If resources are not available to create a thesaurus it is advised that services should use/adapt one in existence. It also states that short-term solutions are more likely to cost more in the long-term.
- *Staff resources:* If staff resources are sparse care should be taken to design a thesaurus that is easily maintained. "This implies that it should be of moderate size and specificity . . . supported by introductory notes and explanations."
- *Buying in resources:* Involves either buying a thesaurus that can be adapted, hiring a consultant, or if buying software, coming up with a list of requirements.
- *Alternatives:* States that a highly complex, structured thesaurus, with "systematic display, specific terminology and compound terms" will be expensive to construct, operate and maintain. "The cost of the effort should be balanced against the expected improvement in performance.

On the other hand, thesauri which are cheap to produce and maintain, having broad terminology and minimum structure, may be more difficult, and therefore more expensive, to operate successfully at the search stage."

Both Chaplan[20] and Whitehead,[21] in the articles sited below, demonstrate that, in terms of labour, manual mapping is time consuming.

Mapping, as undertaken by Whitehead, Chaplan and Busch (1994),[22] shows that there is a large redundancy in terms when mapping vocabularies together. Busch's automatic mapping between the National Art Library (NAL) subject headings and AAT showed that only 51% of NAL heading phrases generated AAT matches. Whitehead suggests that mapping to LCSH is difficult because of its lack of structure. She noted that approximately 16.5% of AAT main terms mapped exactly to LCSH. Chaplan, who reported on the automatic mapping of Laborline Thesaurus to LCSH, said that realistically 41.5% of the terms could be mapped successfully. The approximate time spent on mapping each of the 4,168 Laborline terms was 12 minutes, altogether adding up to 800 hours. Chaplan suggests, that if resources permit, manual mapping, although time consuming, would produce better results.

All three authors stated that where electronic mapping took place a degree of manual intervention was required to aid the process. It is perhaps fair to say, therefore, that if the automated mapping algorithms are robust enough and the methodology is correct then a combination of the two may reduce time and ensure more precision at a high quality.

None of the papers identified gave any indication of financial costs in mapping controlled vocabularies, and it is, in any case, unlikely that costs from other projects would be reliably transferable to the UK scenario that is HILT's primary focus.

GENERAL POSITIVE FEATURES OF A TeRM APPROACH

The advantages of the TeRM based approach are not limited to its possible experimental value; it also has a number of characteristics that make it an attractive facilitator of interoperability in respect of subject terminologies:

- A service of this type should immediately begin to solve the cross-searching and browsing problem to the extent that it should deal with all circumstances where services have used preferred terms from the standard schemes covered by the service. The mapping of UK adaptations and extensions from major services should take this a step further. Retro-conversion of legacy metadata would not be required to ensure this level of improvement, as it would if a single scheme were adopted, for example, and services which already used the standard schemes covered could continue to do so. This would give both the various communities of archives, electronic services, libraries, and museums the flexibility to use schemes they felt were most suited to their users and materials without having to also use an additional scheme to allow cross-searching and browsing. Mapping in older versions of established schemes would circumvent the problem caused by the fact that most services do not retro-convert when new editions of a scheme are produced.

- Basing the approach on internationally recognised schemes and involving international 'players' should help ensure international compatibility.
- Since the approach entails using DDC as one of the schemes, it arguably incorporates most of the advantages of adopting DDC as a single scheme solution,[23] including the multi-lingual advantages of DDC and its widespread use. It is used in over 200,000 libraries and 135 countries world-wide, and is available in over 30 different languages, including those with major world coverage such as English, Spanish, Arabic, and Chinese.
- One clear outcome from the Stakeholder Workshop was that adopting a single scheme was not considered an attractive or acceptable option. This approach circumvents that problem and is likely to be more readily accepted as a result.
- Even the most popular scheme amongst HILT stakeholders–LCSH–is only used by 40% of those responding; 60% would have to change or adopt an additional scheme to allow cross-searching and browsing.
- HILT stakeholders see all schemes as having strengths and weaknesses, with particular communities having particular views on particular schemes.
- Even if a single scheme were adopted, all of the evidence suggests that agencies would adapt and extend it. A TeRM service would provide a focus for registering and recording new terms added to existing schemes and thereby begin to reduce the problems caused, offering staff assigning terms an online overview of those already in use in the UK, together with associated guidance on how they are used, when to use them, and how they relate to terms used elsewhere. It could perform this function, not only for those services that use standard schemes but also for those that currently do not.
- A TeRM service would allow professionals to develop an intelligent overview of subject coverage and inter-relationships in the UK and

would provide an online focus for debate, training, research, and ongoing work.
- It could allow sufficient community control of the terminologies used to ensure compatibility with the requirements of local, regional, sectoral, domain level, and other special interest groups.
- It could offer users an automatic cross-search using equivalent terms from multiple schemes.
- Existing commercial packages already offer some of the flexible interactive tools to assist users and could be developed to offer more.
- It has the potential to improve retrieval by accommodating mappings from the preferred terminology sets of users to the more standard schemes.
- It could offer user training in an online environment–important, given that Web-oriented users seldom access information through intermediaries.
- It is an approach that neither limits an organisation's choice of preferred subject scheme, nor forces it to use an additional scheme to ensure interoperability.
- It provides a process that should, in time, lead to greater harmonisation across domains in the UK, in terms of practice in the field.

THE IMPORTANCE OF CONSENSUS: THE PEOPLE PRINCIPLE

The real significance of the TeRM-based approach proposed, however, is that it would put in place a process allowing the consensus reached within the project to be maintained and built upon, potentially stopping the growth of problems caused by the use of variant schemes and practices more or less immediately, and making the aim of full interoperability in terms of subject terminologies a realisable goal. The importance of this consensus, of the need to maintain it in any follow-up pilot and subsequent service, and to extend it beyond the UK, should not be underestimated. Standards and their consistent application by all may be the bedrock of interoperability, but there is another more fundamental requirement–achieving and maintaining consensus on:

a. Which standards to adopt and how they should be applied.
b. Training for, ensuring, and monitoring adherence.
c. The evolution of the standards and their application to keep pace with changing participant and global requirements.

Without a and b above, interoperability could not be achieved. Without b and c, it could not, in the longer term, be maintained, since changes to the standards and the rules of application would not meet requirements and variant practices would inevitably emerge as a result. Achieving and maintaining 'people interoperability' is thus a prerequisite for achieving and maintaining operational interoperability. This is a difficult task in respect of interoperability generally, but particularly so in respect of subject terminologies. Here, the identification of appropriate descriptive terms often has an interpretative element, the focus of which may change with the purposes of the service or community doing the describing; this is one reason why different practitioners insist on adhering to different subject schemes, and why adopting a single subject scheme as the only standard would, at best, be counterproductive. In these circumstances the HILT proposal, to map schemes in an integrated way in the context of a participatory process that will support and facilitate the maintenance of a consensual approach to ongoing development, arguably offers the best chance of success. In allowing organisations, communities, and domains to retain the use of the subject schemes most appropriate to their requirements, whilst providing an interactive guide to the schemes and their inter-relationships to assist users and facilitate coherent cross-searching or browsing by subject, together with a consensus-based development route towards a more integrated future, the proposal seems to have the potential to offer the best 'safe path' towards future interoperability in respect of subject terminologies in the distributed environment. Assuming that follow-up funding is forthcoming, an initial pilot project should help determine whether or not this potential can be realised in practice, and what level of service is necessary, justifiable, and sustainable in the light of actual user needs and a cost-benefit analysis of different service and mapping levels.

NOTES

1. Gordon Dunsire. "The Internet as a Tool for Cataloguing and Classification: A View from the UK," *Journal of Internet Cataloging*, 2, no. 3/4 (2000): 187-195.

2. <http://hilt.cdlr.strath.ac.uk/>.

3. Strathclyde University's Centre for Digital Library Research (CDLR), UK Office for Library and Information Networking (UKOLN), mda (formerly: Museums Documentation Association), National Council on Archives, National Grid for Learning (NGfL) Scotland, Online Computer Library Center (OCLC), Scottish Library and Information Council (SLIC), and Scottish University for Industry (SUfI).

4. For additional detail about the project and project processes see D. Nicholson, S. Wake, S. Currier, "High-level Thesaurus Project: Investigating the Problem of Subject Cross-searching and Browsing between Communities," in, *Global Digital Library Development in the New Millennium*, ed. Ching-chih Chen (Beijing: Tsinghua University Press, 2001) and Dennis Nicholson and Susannah Wake, "Interoperability in Subject Terminologies: The HILT Project," *New Review of Information Networking*, 7 (2001) (pages not known).

5. See, for example, <http://www.wordmap.com/>.

6. See <http://cairns.lib.strath.ac.uk/> and <http://cairns.lib.gla.ac.uk/>.

7. For a more detailed report see <http://hilt.cdlr.strath.ac.uk/Reports/Focus 2602.html>.

8. See <http://hilt.cdlr.strath.ac.uk/Dissemination/WorkshopNew.html>.

9. Note that this is the actual text used.

10. Full details of the consultation exercise based on the draft Interim Report, including the draft recommendations can be found at <http://hilt.cdlr.strath.ac.uk/Reports/ Consultation.html>.

11. S. Pollitt, "CeDAR: Centre for Database Access Research," *Ariadne*, 4 (1996). Available at <http://www.ariadne.ac.uk/issue4/cedar/intro.html>.

12. H. Gillis, "Guiding the User to Relevant Information Within a Clump, with Particular Focus on an Early Version of the CAIRNS Dynamic Clumping Service," *The New Review of Information and Library Research*, 5 (1999): 99-105.

13. J. Burnett and S. McClure, "Information Services for the Museum of Scotland Project," *Aslib Proceedings*, 46, no. 3 (1994).

14. P. Jascó, "Savvy Searchers Do Ask For Direction," *Online and CD-ROM Review*, 23, no. 2 (1999): 99-102.

15. H. Hug and M. Walser, "Retrieval in the ETH Database Using UDC," in *Tools for Knowledge Organization and the Human Interface: Proceedings, 1st International ISKO-Conference, Darmstadt, 14-17, August* 1990. Advances in Knowledge Organization, vol. 2 (Frankfurt: INDEKS Verlag, 1991).

16. T.E. Doszkocs, "Automatic Vocabulary Mapping in Online Searching," *International Classification*, 10 no. 2 (1983): 78-83.

17. B.R. Schatz, P.A. Cochrane, and H. Chen, "Interactive Term Suggestion for Users of Digital Libraries: Using Subject Thesauri and Co-occurrence Lists for Information Retrieval" (1996). Available: http://dli.grainger.uiuc.edu/papers/schatzDL96. htm.

18. P.A. Cochrane and E.H. Johnson, "Visual Dewey: DDC in a Hypertextual Browser for the Library User," *Advances in Knowledge Organization*, 5 (1996): 95-106.

19. J. Aitchison, A. Gilchrist, and D. Bawden, *Thesaurus Construction and Use: A Practical Manual*. 3rd ed. (London: Aslib, 1997).

20. M.A. Chaplan, "Mapping Laborline Thesaurus Terms to Library of Congress Subject Headings: Implications for Vocabulary Switching," *Library Quarterly*, 65, no. 1 (1995): 39-61.

21. C. Whitehead, "Mapping LCSH into Thesauri: The AAT Model," in *Beyond the Book: Extending MARC for Subject Access*, eds. Toni Petersen & Pat Molholt (Boston: G.K. Hall: 1990): 81-96.

22. J.A. Busch, "Automated Mapping of Topical Subject Headings into Faceted Index Strings using the Art and Architecture Thesaurus as a Machine Readable Dictionary," in *Knowledge Organization and Quality Management. 3rd International Conference*. Advances in Knowledge Organization, vol. 4 (Frankfurt: INDEKS Verlag, 1994).

23. For further information, see <http://hilt.cdlr.strath.ac.uk/Dissemination/Workshop New.html>.

APPENDIX A. Interactive Terminologies Route Map (TeRM) Diagram

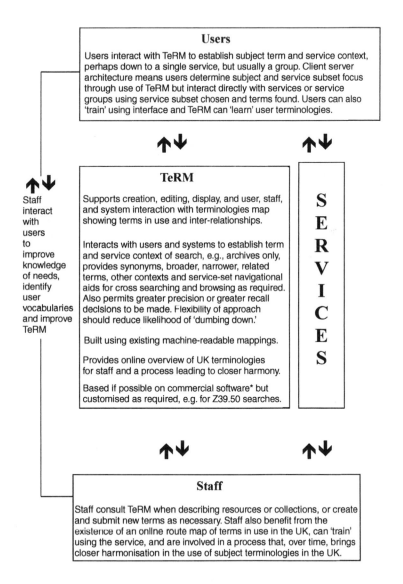

Users

Users interact with TeRM to establish subject term and service context, perhaps down to a single service, but usually a group. Client server architecture means users determine subject and service subset focus through use of TeRM but interact directly with services or service groups using service subset chosen and terms found. Users can also 'train' using interface and TeRM can 'learn' user terminologies.

Staff interact with users to improve knowledge of needs, identify user vocabularies and improve TeRM

TeRM

Supports creation, editing, display, and user, staff, and system interaction with terminologies map showing terms in use and inter-relationships.

Interacts with users and systems to establish term and service context of search, e.g., archives only, provides synonyms, broader, narrower, related terms, other contexts and service-set navigational aids for cross searching and browsing as required. Also permits greater precision or greater recall decisions to be made. Flexibility of approach should reduce likelihood of 'dumbing down.'

Built using existing machine-readable mappings.

Provides online overview of UK terminologies for staff and a process leading to closer harmony.

Based if possible on commercial software* but customised as required, e.g. for Z39.50 searches.

SERVICES

Staff

Staff consult TeRM when describing resources or collections, or create and submit new terms as necessary. Staff also benefit from the existence of an online route map of terms in use in the UK, can 'train' using the service, and are involved in a process that, over time, brings closer harmonisation in the use of subject terminologies in the UK.

*Note: Examples can be seen at <www.wordmap.com> with <www.oingo.com> and <vivisim.com>

Index

AccessER (Access to Electronic
    Resources Task Force), 70-71
  mapping from LC call numbers,
    73-76
Acronyms, 90. *See also* Jargon
Ahronheim, Judith R., 1-3
Automatic subject term germination, 72

BIOME, 6
Bishop, William, 86
Borgmann, Andrew, 82
Browsing services, 2
BUBL LINK, 6

Cataloging, subject
  traditional, 68
  and Web sites, 68-69
Catalogues, 68
Classification schemes, library
  hierarchical, 41-43
  interfaces and, 24-27
  Internet and, 6
  Internet subject trees and, 7-15
Class number mapping, 78-79
Columbia Libraries, 22. *See also*
    Master Metadata File (MMF)
    (Columbia Libraries)
Cross-references, 20

Davis, Stephen, 2,19-45
Dewey Decimal Classification (DDC),
    2. *See also* Library of
    Congress Classification (LCC)
  Internet subject trees and, 8-15

Digital Registry (University of
    Washington Libraries), 52-53
  elements transferring from catalog
    to, 53-55
  lessons learned, 63-64
  maintaining, 59-63
  populating, 55-57
  results, 57-59
  use of, 64-65
Dunshire, Gordon, 3,97-110

Electronic journals, 70
Electronic lists, 69-70
Electronic resources
  access methods for, 52
  classification, 6
Enterprise Information Portal (EIP), 84

Forsythe, Kathleen, 2,51-65

Google, 43
Graphical browsing techniques, 12

Hierarchical classification systems, 41-43
High-level browsing vocabularies, 2
HILCC (Hierarchical Interface to Library
    of Congress Classification), 19.
    See also Master Metadata File
    (MMF) (Columbia Libraries)
  assessing, 39-40
  implementation issues, 37-39
  Phase II for, 39-41
  pilot project for, 27-28

project planning for, 27-28
  working design principles and
      considerations for, 28-37
HILT (High-Level Thesaurus) project,
      2,98-99
  interim report, 102-103
  problems of, 103-106
  stakeholder workshop for, 100-102
Hummingbird, 84

Information, factors required to
      produce, 82
Information Gateway, 52-53,54. *See
      also* Digital Registry
      (University of Washington
      Libraries)
Information overload, 82
  university libraries and, 83
Interfaces
  LC classification as basis for, 24-27
  OPACs and, 21
Internet, library classification and, 6

Jargon, 87,90

Keyword searches, 43

LibQUAL+, 83
Libraries, university, and information
      overload, 83
Library classification schemes. *See*
      Classification schemes,
      library
Library of Congress Classification
      (LCC), 2. *See also* Dewey
      Decimal Classification (DDC)
  hierarchical interfaces to, 43-44
  Master Metadata File (MMF)
      (Columbia Libraries) and,
      24-27

Library portals, 84-86,89-90
Lists, electronic, 69-70
Lynch, Clifford, 87

McDonalization, 88-89
Mapping, 3,14,73
  exhaustivity of class number, 78-79
  from LC call numbers, 73-76
Master Metadata File (MMF)
      (Columbia Libraries), 22-24.
      See also HILCC
      (Hierarchical Interface to
      Library of Congress
      Classification)
  LC classification and, 24-27
Morgan, Keith A., 2,81-95
MS-SQL, 53
MyLibrary, 90-93

Neill, Susannah, 3,97-110
Nicholson, Dennis, 3,97-110
Number mapping, class, 78-79

OPACs (Online Public Access
      Catalogs), 20-21
  information retrieval and, 20
  interfaces and, 21
Ormsby, Eric, 86,87

Portals, 84-86,89-90
Proliferating subject schemes, 71-72

Quinn, Brian, 88-89

Reade, Tripp, 2,81-95
Renardus project, 6-7,12
Research, Internet and, 87-88

Risen, Kirsten, 2
Ritzer, George, 88,89
Rothman, Jonathan, 2,67-80

Shadle, Steve, 2,51-65
Smith, Frank, 89
Subject cataloging
   traditional, 68
   for Web sites, 68-69
Subject heading searches, 43
Subject schemes, proliferating, 71-72
Subject term generation, automatic, 72
Subject trees, 2
   library classification schemes and,
      7-15
Task-based subject lists, 2
TeRM (terminologies route map), 99
   advantages of, 106-108
   importance of consensus and, 108-109
Topic maps, 3,77-78

University libraries, and information
      overload, 83
University of Michigan Libraries,
      69-71

University of Washington. *See* Digital
      Registry (University of
      Washington Libraries)
URL maintenance, 53
Usability testing, 90
User-focused subject lists, 2

Vizine-Goetz, Diane, 2
Vocabularies
   high-level browsing, 2
   mismatches in, 68

Washington, University of. *See* Digital
      Registry (University of
      Washington Libraries)
Web portals. *See* Portals
Web sites, methods of categorizing,
      68-69
Wheatley, Alan, 2,7
WorldCat, 2,6

Yahoo!, 41-43,68-69